ROSA

ROSA

the story of the rose

PETER E. KUKIELSKI

WITH CHARLES PHILLIPS

FOREWORD BY JUDITH B. TANKARD

Yale

UNIVERSITY PRESS

New Haven and London

Yale University Press books may be purchased in quantity for educational, business, or promotional use. For information, please e-mail sales.press@yale.edu (U.S. office) or sales@yaleup.co.uk (U.K. office).

This book has been composed in Effra and Bell.

Printed on acid-free paper.

The Bright Press, an imprint of the Quarto Group,
The Old Brewery, 6 Blundell Street,
London N7 9BH, United Kingdom.
T (0)20 7700 6700
www.QuartoKnows.com

Publisher: James Evans
Editorial Director: Isheeta Mustafi
Art Director: Katherine Radcliffe
Managing Editor: Jacqui Sayers
Commissioning Editor: Sorrel Wood
Development Editor: Abbie Sharman
Project Manager: Kate Duffy
Design: JC Lanaway

Printed in Singapore by C.O.S. Printers Pte Ltd.

Library of Congress Control Number: 2020940863

ISBN: 978-0-300-25111-1 (hardcover: alk. paper)

A catalog record for this book is available from the British Library.

10 9 8 7 6 5 4 3 2 1

CONTENTS

FOREWORD

Judith B. Tankard

There's no question that for centuries roses have captured the imagination of almost everyone, from a welcome florist's bouquet to the scientific development and naming of new varieties by growers. Where would political history and intrigue be without roses? Or for that matter, can one imagine a traditional, English garden without roses? During the Edwardian age, there was scarcely a garden painting without cascades of roses. When Alfred Parsons, who was one of the best illustrators of that era, produced *The Genus Rosa* for Ellen Willmott of Warley Place in 1910, it rivaled Marie Antoinette's famous volumes. The horticulturalist and writer Gertrude Jekyll, who shared many words of wisdom in her books, reached thousands of eager readers both British and American in her practical book, *Roses for English Gardens* (1902), based on years of experience. Years later Beatrix Farrand, the well-known American landscape gardener (as she called herself), was an admirer of Jekyll as well as of roses. Farrand once quipped, "Don't plant a rose garden unless you mean to look after it." As a child she learned about roses from her grandmother whose extensive gardens in Newport were filled with roses. Farrand's practical knowledge of roses proved essential when she was commissioned in 1915 to design a large rose display garden at the New York Botanical Garden (now called the Peggy Rockefeller Rose Garden). For inspiration she modeled the garden after the Roseraie de l'Haÿ (now Roseraie du Val-de-Marne) in France, which has collections of over 2,000 varieties.

Nearly 100 years later it was in Farrand's rose garden at the New York Botanical Garden that I first met Peter Kukielski. His work in revitalizing Farrand's original garden with new varieties was truly inspired and I am sure that both Jekyll and Farrand would have been awed at the collections. Peter's readable and engaging volume, an undertaking of many years, is sure to inspire new generations of rose lovers.

The Roseraie du Val-de-Marne in France is a garden
dedicated to roses. It is renowned for its use of
architectural elements and decorative structures to
display the many thousands of different rose varieties
that can be found in this unique garden.

PREFACE:
STORIES OF ROSES

Won't you come into the garden?
I would like my roses to see you.

SHERIDAN (1751–1816)

The Greek philosopher Aristotle is credited with the phrase: "The whole is more than the sum of its parts" (c. 350 BCE). In regard to this book, we will define "the parts" as those stories told with amazing, inspiring, and resilient roses. The full account of the rose is a continuing one—of which we detail the beginning as with any other writing, with Chapter 1. The first chapter always sets the stage and introduces the beginning characters so the rest of the book can follow. Here, we start with a "Once upon a time... some 35 million years ago in a land far away" kind of opening.

Within this book, the roses are the characters in these stories. The chapters that follow reveal numerous scenes in which roses play: archaeological studies, myths, pleasure gardens, and ancient cultures. Roses perform in religions, love stories, poetry, literature, and wars. The rose influences art and architecture; and shapes fragrance and medicine.

In Greek and Roman times, the rose gains a prominent place in society. Through myths, Aphrodite and Venus share the rose; and Theophrastus, Sappho, and Pliny all bring importance to the rose through their writings. Alexander the Great, a huge rose lover, captures the bloom among quests;

and Nero and Heliogabalus (Elagabalus) display some rose mania. Symbolically through Christian times, the rose takes a pew with the Virgin Mary, the rosary, and saints. Later, a famous balcony scene depicts Shakespeare's claim that "a rose by any other name would smell as sweet" (*Romeo and Juliet*), and artists such as Renoir, van Gogh, and Manet all found inspiration from the rose as revealed in many of their paintings. Islam and Sufism have the rose where we find Rumi, who inspirationally expresses it in his famous *Masnavi*, possibly one of the greatest masterpieces of poetry.

Like legendary performances we know today, some leading ladies in these chapters include the great roses of their era: *Rosa gallica*, *Rosa sancta*, 'Autumn Damask,' and 'Kazanlik.' Other roses carry the specific name association of their famous counterparts, such as 'Souvenir de la Malmaison'—a light-pink Bourbon rose that is undeniably matched with Empress Joséphine at her famous chateau. The sixty-petaled rose commonly called "King Midas" (*Rosa damascena*), see Chapter 3, is very distinct from the five-petaled species roses found in China regions, see Chapter 1. Both of these are unique to the hundred-petaled *Rosa centifolia*, which was painted by late seventeenth- and eighteenth-century artists, including Rachel Ruysch and Pierre-Joseph Redouté, as discussed in Chapters 7 and 8. Furthermore, one wonders what the real rose was that Cleopatra used to woo and seduce Mark Antony at the height of the Roman Empire.

In religious symbolism, we find the rose garden as a standard image of paradise. With health and medicines, rose water and rose oil have centuries-old uses, and entire societies were created around these industries. Roses were exchanged as currency. The traveling rose has moved from China, across Asia and Persia, to Ancient Europe, Egypt, and finally following trade routes around the world. Moreover, the fragrance of the rose is undeniable. It gives us wonder and is spirit-lifting; its aroma can calm the nerves and rest the soul.

In this book, you can read about particular roses and the roles they played through history, see the various lands from which they came, and begin to understand the essential contributions of all of them. Enjoy the stories, the symbolism, and the joy that they bring.

A rose is a rose is a rose.

Sacred Emily, Gertrude Stein, 1913

Rosa damascena subalba

1
ROSA FAMILY OF PLANTS

When inspiring ideas about roses, Gertrude Stein's poem *Sacred Emily* contains the phrase "A rose is a rose is a rose." Here, Ms. Stein metaphorically indicates that "it is what it is," expressing the fact that merely using the name of something already invokes the imagery and emotions associated with it. However, if people believe this statement, they are missing out on the gorgeousness of the individuality of every rose. One of the most beautiful things about roses is learning and experiencing their different attributes: their fragrances, their colors, their growth habits, and the flower forms within their various families. It is also interesting to understand that the rose has evolved and changed throughout its existence.

BOTANICAL NOMENCLATURE AND THE ROSE

THE ROSE is part of a larger group of plants in the Rosaceae family (the term "Rosaceae" refers to a family of flowering plants). Numbers can vary, based on the source, but this family includes between 3,500 and 4,800 species and approximately 100–160 genera. The family Rosaceae includes some shrubs, trees, and herbaceous plants, as well as roses. This plant family exists because of the commonality of their bloom.

Botanical nomenclature is the scientific naming of plants. The naturalist Carl Linnaeus (1707–1778), is credited with the starting point of this system. Before Linnaeus, botanical nomenclature has a long history from when Latin was a primary language to Theophrastus and Pliny the Elder. Horticulturists and botanists use this system to identify plants by name. "A rose is a rose is a rose" isn't adequate. However, the name *Rosa × damascena* 'Quatre Saisons' tells us the rose is pink, a larger grower, and because it is a Damask, deliciously fragrant. We also understand that the further qualifier 'Quatre Saisons' indicates that this particular rose is the unique Damask with some sporadic blooms in the fall—the 'Autumn Damask.' With this one name, we can paint the accurate picture of what this plant looks and smells like, as opposed to just saying "pink" roses.

I recall being a frequent guest on a popular talk radio show out of New York City. Listeners from around the country would call in with their questions about roses. Without fail, every time, I would get the caller(s) who would ask why their pink rose is having

problems. I would respond by begging, "Can you tell me any more about your rose, other than that it is pink?" If the caller could tell me, for example, that their 'Queen Elizabeth' rose (which is pink) is having issues, then because the rose's name was supplied I would know how to help them. A pink rose can generally be a rose, but more specifically, the 'Queen Elizabeth' rose is a tall and upright Grandiflora producing clusters of pink blooms, and is moderately fragrant. The name told me how their rose was supposed to grow and its exact color. I could tell if it was disease-susceptible or even hardy, depending on which part of the country they lived in, and I could therefore help them with their questions.

THE ROSACEAE FAMILY

Some of the groups of plants under the Rosaceae family umbrella include:

Sorbus Widely known as mountain ash, rowan, service tree, and whitebeam.

Crataegus Commonly known as hawberry, hawthorn, May tree, quickthorn, and whitethorn.

Rubus Many commonly known members are berries: blackberry, boysenberry, loganberry, raspberry, and tayberry.

Prunus Commonly known members are almonds, apricots, cherries, peaches, and plums.

Fragaria Commonly known as strawberry.

Malus Commonly known as apple.

'Queen Elizabeth' rose

Species Roses

Species roses (those that nature gave to us millions of years ago) have a characteristic flower form that is flat and only has five petals. We call this a simple flower form. One exception is *Rosa sericea pteracantha*, which has four petals. Numerous other plants also have five-petaled flowers.

The common aspect of these roses is the five-petaled flower. This trait places them all under the same plant family—Rosaceae (see page 13). One other subset of Rosaceae, other than those outlined, is the genus focus for this book: *Rosa*.

The genus *Rosa* itself has four subdivisions (subgenera), each of which has identifying characteristics that place them in the following subgroups.

Hesperrhodos: Desert rose (*Rosa stellata*)

Hesperrhodos

From the Greek for "western rose," this genus includes two species from southwestern North America: *Rosa minutifolia* and *Rosa stellata*.

Hulthemia

Rosa persica and *Rosa berberifolia* are native to Persia as sturdy, thorny plants that bloom sparsely. They are wild and can be found growing in the desert-like conditions of Iran and Afghanistan. They have a unique bloom with a center "eye" that is distinguishable from other flowers.

Hulthemia: Barberry-leaf rose (*Rosa persica*)

Platyrhodon: Chestnut rose
(*Rosa roxburghii* f. *normalis*)

Platyrhodon
From the Greek for "flaky rose," there is one species from East Asia called *Rosa roxburghii* f. *normalis*.

Rosa
Containing all other roses, subdivided further into eleven sections.

Native to China
1. *Banksianae*: Native to central and western China.
2. *Bracteatae*: Also found in India, Japan, Taiwan, and the Himalayas.
3. *Chinensis*: Native to southwest China in Guizhou, Hubei, and Sichuan Provinces, and also India and Myanmar (Burma).
4. *Laevigatae*: Also found in Taiwan, and south to Laos and Vietnam.

Native to Europe
5. *Caninae*: Also native to northwest Africa and western Asia.
6. *Gallicanae*: Also found in western Asia.
7. *Pimpinellifoliae* ('Burnet rose'): Also native to northwest Africa and Asia.

Native to North America
8. *Carolinae* ('Carolina rose'): Found in eastern North America.
9. *Gymnocarpae* ('Dwarf rose,' 'Baldhip rose,' 'Wood rose'): Native to western North America and Asia.

Native to all regions of the Northern Hemisphere
10. *Synstylae* (*Rosa multiflora*): Native to eastern Asia, China, and Japan.
11. *Cinnamomeae*: Except northern Africa. These roses further subdivide into *Rosa rugosa*, *Rosa arkansana*, *Rosa blanda*, *Rosa pendula*, *Rosa oxyodon*, *Rosa laxa*, and *Rosa majalis*.

Rose Groupings

There are three main groupings of roses under the umbrella of *Rosa*: species, old roses, and modern roses.

In the rose world, there is a significant date of division. 'La France' was introduced in 1867 by Jean-Baptiste Guillot (1827–1893) by crossing a Hybrid Perpetual and a Tea Rose. 'La France' is generally accepted to be the first Hybrid Tea Rose in Europe.

All roses before 1867 are considered "old" roses, and all the roses dated after 1867 are classed as "modern." Since this book tells the stories of roses through history, going back 35 million years to the present day, the majority of roses in this book falls into the species category or the old rose category. The vast majority of rose expansion exists in our last century to modern day, as roses began to be cultivated in exponential numbers (these modern roses don't appear until Chapter 9). Today, rose societies define 37 unique classes of roses. Each rose class is based upon its common flower form unique within them.

Rosa centifolia

'La France'

Old Rose Classes

- Alba
- Ayrshire
- Bourbon and Climbing Bourbon
- Boursault
- Centifolia
- China and Climbing China
- Damask
- Hybrid Bracteata
- Hybrid China and
  Climbing Hybrid China
- Hybrid Eglanteria
- Hybrid Foetida
- Hybrid Gallica
- Hybrid Gigantea
- Hybrid Multiflora
- Hybrid Perpetual and
  Climbing Hybrid Perpetual
- Hybrid Sempervirens
- Hybrid Setigera
- Hybrid Spinosissima
- Hybrid Stellata
- Miscellaneous Old Garden Roses
- Moss
- Noisette
- Portland
- Tea and Climbing Tea

Modern Classes

- Floribunda and Climbing Floribunda
- Grandiflora and Climbing Grandiflora
- Hybrid Kordesii
- Hybrid Moyesii
- Hybrid Musk
- Hybrid Rugosa
- Hybrid Wichurana
- Hybrid Tea and Climbing Hybrid Tea
- Large-flowered Climber
- Miniature and Climbing Miniature
- Miniflora
- Polyantha and Climbing Polyantha
- Shrub

Climbing rose, 'New Dawn'

PRICKLES AND HIPS

Rose hips are the fruit of the rose, which contain its seed. They are rich in vitamin C and used for cooking and many medicinal remedies. The open-faced flowers of species roses are attractive to pollinating bees and other insects, thus are more apt to produce hips than the tighter, many-petaled flowers. There can be as many unique and colorful hips as there are roses, such as *Rosa rugosa* and *Rosa moyesii*.

During winter months, the hips are a colorful interest in the landscape and provide food for birds and animals. When we look (or touch the stem of the rose), we unfortunately may encounter the sharp objects appearing along the cane. Surprisingly, these are not thorns—the proper term for these is prickles! Prickles are outgrowths of the outer layer of tissue of the stem and provide protection. Prickles have a hook shape, which aids in climbing.

Rosa rugosa rose hips

Classes of Roses

Let's look in more detail at a few of the significant classes of roses that you would see in garden centers and specialty rose nurseries today.

Species/Wild Roses

These are the roses that nature has given us—the wild, naturally occurring roses of this world. They bloom once a year (called "once-blooming") and usually have a single flower form. Often a subject of debate, there are more than 100–150 (up to 300) exact species. All roses exist only in the northern hemisphere and in temperate regions, and many species of roses are known for their rose-hip displays.

Old Roses, Old Garden Roses, Antique (before 1867)

Hybrid Gallica

Rosa gallica is a species rose native to southern and central Europe and has been known for thousands of years. The earliest named form of this plant is *Rosa gallica* 'Officinalis' (also known as the apothecary's rose because it was thought to have medicinal properties). *Rosa gallica* 'Officinalis' has multiple petals—as do other descendants of

Rosa gallica 'Officinalis'

Rosa gallica—sometimes up to one hundred petals per flower. Other Gallicas in this class are hybrids of the species and display shapely, highly scented flowers, often in saturated deep colors.

Damask rose, 'Oeillet Parfait'

Alba rose, 'Great Maiden's Blush'

Damask

The original Damask rose was thought to be a cross between *Rosa gallica* and another species of rose, but many experts argue over this point. Damask roses do not have as many petals as the Gallicas before them, yet they have many more petals than the earlier species. The Damasks are mostly known for their fragrance and are the primary source for rose oil or "attar of roses," which has many uses, including a source for modern-day perfumes. They are once-blooming, except for 'Autumn Damask' (see pages 65 and 90).

Alba

Alba means white, but the Albas include roses that range from whites to pinks. Albas are understood to be a cross between *Rosa canina* and *Rosa damascena*. They have a wider variety of flower forms and a lighter fragrance than the heavy damask scent of their predecessor. Another characteristic of this class is that the foliage shares a glaucous (grayish-green or blue) coloring, which differs from the brighter green foliage of earlier types, and it is once-blooming. Albas were also known as "tree roses," as they can grow to six feet (1.8 meters) in height.

Centifolia

A simple qualifying characteristic of Centifolia roses can be gleaned from the name "centi," meaning hundred. These blooms, which are quite different from other classes of roses, are generally very cupped or rounded in shape and do indeed have about one hundred petals. Centifolia roses (also known as "cabbage roses" for their flower form) are said to have been the first class developed as late as the seventeenth century. Maybe because of this later date of cultivation, the number of Centifolia roses in commerce appears to be fewer than in other classes.

Moss

Moss roses have bloom characteristics similar to those of their predecessors, therefore they can be difficult to identify simply by looking at their flower form. The apparent difference is the addition of "mossy" glands to the buds. These are covered with glandular tips that produce their own oily scent, ranging anywhere from citrus and anise to earthy notes, which make an excellent complement and contrast to the rose scent of the flowers—two different perfumes from the one plant! Moss roses are known to be sport (see page 23) descendants of the Centifolias.

Centifolia rose, 'Dometil Beccard'

Moss rose, 'Comtesse de Murinais'

Hybrid China

China roses, somewhat obviously, come from China, and they were developed from the species *Rosa chinensis*. China roses are among the most floriferous of roses and are credited with having given most modern roses their remontant (reblooming) qualities. They also introduced new flower forms and colors to cross with existing European roses. The China roses include shades from white to pink. They also brought some yellow to pastel shades, darker reds to crimsons, as well as purple colors to the rose world. Any rose called a Hybrid China would be a cross between the China roses and other existing classes of rose (see page 17).

Portland (Damask Perpetual)

Because roses inherit the genetic attributes of all their ancestors, the number of genetic and possible hybrid crosses available throughout the world of roses grows exponentially. It can be difficult to tell certain classes apart because they share so many heritable characteristics. Portland roses offer the repeat-blooming traits that come from their China rose heritage. They provide some of the same classic flower forms as their previous parent classes of Gallicas and Damasks. One unique characteristic of Portland roses is that their bloom is "presented" with a grouping of leaves just below the flower, much like a nosegay.

Hybrid China rose, 'Archduke Charles'

Portland rose, 'Comte de Chambord'

PLANT SPORTS

I'm not talking about baseball here. In the botanical world, a sport is part of a plant that shows differences from the original plant. This difference is generally thought to be a genetic mutation. Here are a few examples:

- A different growth habit, i.e., a shrub form of a plant suddenly showing a more elongated cane and climbing form.

- A plant that typically blooms once, but unexpectedly has a repeat bloom.

- A plant that usually has one flower color, yet suddenly displays a different color bloom, i.e., from white to pink.

- Moss roses resemble the Centifolia roses, except that their buds are covered by glands that result from a genetic mutation.

Bourbon

With roots that link back to the China and the Damask roses, Bourbons have the qualities of both. From their China ancestors, the Bourbons seem to have gained a broad range of colors, a delicacy in their blooms, and good reblooming ability. From the Damasks, the Bourbons show some flower-form characteristics and deep fragrance. Bourbons are very easy to find in commercial nurseries and are lovely plants for the scented garden.

Bourbon rose, 'Louise Odier'

Noisette rose, 'Alister Stella Gray'

Hybrid Perpetual rose, 'Paul Neyron'

Noisette

Noisettes are some of the great flower producers in the rose garden. They are generally a more southern plant as they are not very winter hardy. The genes for powerhouse blooms come from their China parentage, and their pleasant fragrance and clustering of flowers come from the species *Rosa moschata*. The Noisettes date back to 1802, when John Champneys of South Carolina developed some seedlings; the brother of Champneys' neighbor, French nurseryman Philippe Noisette, further developed them in France.

Hybrid Perpetual

Hybrid Perpetual roses are a combination of all the rose genetics before them. Within the complexity of this class, a new flower form emerged that tended to be rather large, full, and higher-centered than previous classes. This voluptuous bloom got the attention of breeders and customers alike, and people began to make crosses of roses more than ever before.

Tea

Tea roses are repeat-flowering roses with a fragrance resembling that of the black tea that we drink. It is more of a southern climate plant.

Modern Roses: In Existence Since 1867

Hybrid Tea

These are arguably the most numerous roses, and many new varieties enter the market each year. They are repeat-blooming, free-branching shrub roses of upright or bushy habit, with prickly stems and glossy or matte mid-dark green leaves. They have one large—usually double—sometimes scented bloom per stem. The desire for these flowers even spurred a new industry: production and sale of the florists' rose. These flowers are unlikely to have any fragrance, however, newer hybridization efforts are successfully bringing the scent back.

Polyantha

Typically, compact shrub roses with prickly stems and glossy green leaves, Polyanthas are usually repeat blooming. Sprays of small, single to double flowers are produced in flushes from late spring until the fall.

Miniature and Miniflora

These are compact shrub roses with very short stems and small flowers and leaves. The blooms are repeat blooming and may be produced singly or in small clusters.

Hybrid Tea rose, 'Francis Meilland'

Grandiflora

This class results in a cross between a Hybrid Tea and a Floribunda, and therefore shares the characteristics of both. They are repeat-blooming, free-branching shrub roses of upright or bushy habit, with prickly stems and glossy or matte mid-dark green leaves. Large, usually double, often scented flowers are generally solitary, although sometimes produced in clusters. The first Grandiflora was the 'Queen Elizabeth' rose, introduced in 1954 (see page 13).

Floribunda rose, 'Poseidon'

Shrub rose, 'Carefree Beauty'

Floribunda

The Floribunda class resulted from crossing a Hybrid Tea with a Polyantha. Floribundas are free-blooming, free-branching shrub roses of upright or bushy habit, usually with prickly stems and glossy green leaves. Single to fully double, sometimes scented flowers grow in clusters of three to twenty-five.

Shrub

The Shrub class is a diverse, catchall group, where the plants are usually larger than Hybrid Tea roses. The flowers are often scented, single to fully double, mostly remontant, and bloom in clusters from late spring to fall. Often roses are hybridized to the extent that they become their very own class of roses. The beautiful David Austin English Rose collection is an example.

Large-flowered Climber

These are vigorous climbing roses with arching, stiff canes that are typically covered with prickly thorns, and often dense, glossy green foliage. Many have scented flowers in a variety of forms, borne singly or in clusters. Some bloom in spring or early summer, only on short shoots that emerge from the previous year's canes; others are repeat-blooming and flower on new canes. The 'Nahema' rose is an example (see below).

Hybrid Musk

These roses are probably from the species *Rosa multiflora*, providing its abundance of blooms and *Rosa moschata*, giving its musk fragrance. These are vigorous, spreading shrubs that bear graceful blooms in clusters. The flowers are highly scented and repeat all season long. These roses are best when grown freely to let their natural shape develop in the landscape.

Hybrid Rugosa

These roses are vigorous, hardy shrub roses with wrinkled ("rugose" means wrinkled or corrugated) and usually bright green leaves. *Rosa rugosa* 'Alba' and *Rosa rugosa* 'Rubra' are forms of the species. These roses have been cultivated so much that they have formed their own class. Most bear clusters of single or semi-double, scented flowers. Rugosas also produce lovely, large, and colorful rose hips.

Large-flowered Climber rose, 'Nahema'

Hybrid Rugosa rose, 'Agnes'

Come, the rose garden

has flowered!

Teachings of Rumi, AFLAKI, 1999

Rosa palustris Marsha
(formerly *Rosa*
Hudsonian Salifolicea

2

THE ANCIENT ROSE STORY

The rose induces passion, is symbolic of many
things in many cultures, and has an ancient story that
chronicles its strength and endurance, which proves
its legend. We need to explore a little of the history of
the Earth in a geological sense, which will help us to
understand the "ancientness" of the rose and shed light
on how it has survived and endured over the millennia.

The fact that rose species have determined a way
to survive and thrive on earth through 35 million years
of major geological transitions is a fantastic achievement.
Through continents crashing together, mountains rising,
and ice caps forming, the rose survives. It proves that
it is one genetically robust plant.

THE ORIGIN OF THE ROSE

THE EARLY HISTORY of the rose is a mystery. The leaves in most rose species are challenging to identify as they are chartaceous (paper-like), so the thin, fragile tissue breaks down readily and would not be easily fossilized. Therefore, we do not have a reliable recorded history of the rose for millions of years. However, it is thought that the rose may have first appeared in central Asia, and fossils dating from 35 million years ago give paleontologists evidence of roses at that time.

The Eocene Epoch (56–33.9 million years ago)

From the Greek word meaning "dawn," the Eocene epoch is characterized by the appearance and diversification of new groups of organisms. By this time, the supercontinent Pangaea had been broken apart to form two vast landmasses that drifted, Laurasia became the northern continent and Gondwana the southern continent.

Warm equatorial currents mixed with the colder Antarctic waters, distributed heat around the planet, and kept global temperatures high. Then, the split of Australia from the southern continent routed warm currents northward. As a result, Antarctica began to freeze.

RIGHT During the Eocene the climate was far warmer and more humid, allowing for lush vegetation to flourish in regions such as modern-day Greenland and Alaska.

Further break-up of Laurasia and Gondwanaland resulted in the formation of continents drifting toward their present positions. The creation of mountains in modern-day North America began, and vast lakes formed in the high, flat basins. Though the landmasses were shifting, the flora and fauna of the lands suggest there was still a land connection. As the subcontinent of India moved north it collided with Asia with such impact that the Himalayas were formed.

Conditions during the Eocene meant that extensive forests formed in what are today snow-covered polar regions. Fossils confirm that swamp cypress and dawn redwood trees grew on what is now the Arctic island of Ellesmere. There is evidence too from the fossil record in both Greenland and Alaska of subtropical and tropical trees.

The warm temperatures and humid conditions meant that subtropical forests were common to both North America and Europe. The growing conditions during the Eocene in each of these locations were favorable to roses as well, which benefit from six to eight hours of sunlight per day, and typically grow in temperate to subtropical climates with moist (but not wet) soils. Although promising, physical evidence of roses does not exist during this epoch. Later, the geology of the Oligocene epoch (see page 34) reveals rose fossils for the first time.

56 million years ago

Laurasia

North America   Europe   Asia

Africa

India

South America   Gondwana

Australia

Antarctica

The Study of Rose Fossils from the Eocene Epoch

A renowned paleobotanist, Herman Becker, studied rose species from their fossils. In his study Mr. Becker found that:

> ... *fossil floras, containing roses, are represented by species based on only one or very few leaves or leaflets.*
>
> The Fossil Record of the Genus Rosa, *Bulletin of the Torrey Botanical Club*, 1963

The conclusion is that it was impossible to correlate the fossil species to "any one living plant," as no quantitative evaluations were possible. Becker concludes that it is hard to prove one variety of rose over another as the leaves (growth habit) of plants vary according to their location and the climate. If more rainfall occurs in one area, the leaves of the plant will have the propensity to be larger than the leaves of plants that receive little to no rain, which explains the difficulty of identification of the different varieties of ancient roses.

Rose fossils from the Eocene epoch reveal themselves in North America at Florissant, a small village in South Park, Colorado, west of Colorado Springs, and at Bridge Creek and Crooked River in central Oregon. The rose fossil specimens collected at these two sites have been named *Rosa hilliae, Rosa wilmattae, Rosa scudderi*, and *Rosa ruskiniana*. The living roses they most closely resemble today are *Rosa nutkana* and *Rosa palustris*. Although four separate fossil species are cataloged here, they may originate from a single species. But, they have not been identified with complete accuracy due to a shortage of rose curators and lack of funding; new, technological advances have only become helpful in recent years. These fossils dated back 35 million years ago. It is truly amazing to think of the "ancientness" of the rose.

LEFT *Rosa palustris* Marshall, also known as the "swamp rose," is a very fragrant species rose that favors wet ground, such as swamps, marshes, and stream banks. This capacity to flourish in saturated conditions makes it an oddity among roses.

ROSE FOSSILS

Plant fossils form when a plant dies in a watery environment and is buried in mud and silt. Soft tissues quickly decompose, leaving behind twigs and harder plant material. Sediment builds over the top and hardens into rock.

Unfortunately, flowers are often made up of soft tissue and are likely to decay before they can become fossils. Flowers are often the most reliable means of identification, particularly for roses, which makes rose fossils challenging to identify and, therefore, very rare. There are far fewer fossils of roses than other plants, which suggests that roses existed mostly in drier places than plants that require a higher than average rainfall, resulting in a significantly reduced chance of fossilization taking place.

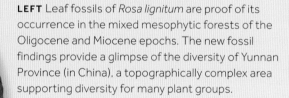

LEFT Leaf fossils of *Rosa lignitum* are proof of its occurrence in the mixed mesophytic forests of the Oligocene and Miocene epochs. The new fossil findings provide a glimpse of the diversity of Yunnan Province (in China), a topographically complex area supporting diversity for many plant groups.

The Oligocene Epoch (33.9–23 million years ago)

The Oligocene epoch marks a significant change between the tropical Eocene that predates it and the cooler Miocene epoch (23.03–5.333 million years ago) that ensued. Significant shifts during this time introduced major expansion of grasslands. This change resulted in the forests retreating to areas around the equator. A famous destruction episode associated with this climate change (cooling temperatures) is called the Grande Coupure. This period is the start of the Oligocene epoch. Asian fauna replaced the now extinct European fauna; western Europe witnessed the introduction of many new taxa, including artiodactyls and perissodactyls (hoofed animals) from areas to the east, and many Eocene genera and species became extinct.

The Oligocene Climate Change

Climate change during the Oligocene epoch saw the earth experience a rise in ice volume and a 181-foot (55-meter) decrease in sea level—associated with a decline in temperature. At this time, the continents began to assume their present positions. What was to become Antarctica shifted farther away from the equator and formed an ice cover. In modern-day North America, the western mountain ranges continued to develop, while in Europe, the Alps started to emerge when the African and Eurasian continental plates collided.

Flowering plants (angiosperms) continued to spread throughout the world as temperate deciduous forests replaced tropical and subtropical forests. More plains and deserts formed on the planet and grasses took root in these open plains. In North America, subtropical species—roses, cashews, and lychees—dominated, and temperate trees such as beeches and pines were present. These changes resulted in ecosystems that flourished in a cooler climate during the Miocene epoch.

BELOW The cooler climate of the Miocene epoch caused the evolution of fauna and flora adapted to more drier, more open, conditions.

THE CHINA IMMERSION

SOUTHWESTERN CHINA is a biodiversity hotspot, and many garden plants develop from wild ancestors native to this region, for example: the azalea, the camellia, the lily, and the rose. China is presently the center of diversification for roses. There is an average of approximately 160 existing species widely distributed from temperate and subtropical regions to very cold regions. Among these, some ninety-five species are native to China, of which around sixty-five are endemic.

This region is known as the "Mother of (rose) Gardens," a fitting description given by various horticulturists. Among these garden plants, the rose is well known for its showy flowers and fragrance. Wild, naturally occurring species roses in this region include *Rosa chinensis*, *Rosa gigantea*, *Rosa moschata*, *Rosa multiflora*, and *Rosa wichuraiana*. These, along with the European *Rosa gallica* and *Rosa foetida*, found in the Caucasus Mountains, have made the most contributions to *Rosa* cultivars.

Chinese Fossil Records

Luckily, reports of rose fossil records with leaves "intact" and well-preserved come from late Miocene deposits in Yunnan Province in southwestern China. As previously mentioned, roses are difficult to identify in fossil form because most species have chartaceous leaves and they decay quickly, making them less likely to be preserved as fossils. Additionally, the rose is never dominant in the modern vegetation of subtropical regions; rose species are

LEFT *Rosa foetida*—the word *foetida* means "having a bad odor or smell." Although the fragrance is considered unpleasant by some, yellow *Rosa foetida* was an essential component of historical rose breeding.

usually restricted to the openings in the canopy or along forest edges. One example, *Rosa fortuita*, is a new rose species, identified in 2015 from a fossil record with good leaf preservation. The discovery of this rose provides evidence that roses existed in southwestern China at least by the late Miocene and it may have occupied the same ecological niche as present-day species.

Rose Fossil Discoveries in the Dashidong Area

In the Dashidong area of Yunnan Province, finding only three specimens of roses among more than 2,000 fossils discovered there proves again that roses are rare in fossil form. The rose fossil discoveries in Dashidong suggest that roses grew in temperate to subtropical forests. For example, *Rosa fortuita* grew in a mixed mesophytic forest, a forest that generally receives a moderate amount of moisture, with a subtropical or temperate climate. Existing rose species in Yunnan Province support this theory. The diverse altitudes of 5,900–11,200 feet (1,800–3,400 meters) offer subtropical or temperate (mesophytic) climates. The general climate requirements of roses in southwestern China appear to have remained the same since the Miocene epoch, which would explain the breadth of genetic material there and why China is considered a principal source of rose genetics—a real jackpot in a rose genetics sense. (It's like a kid in a candy store!)

Climatic and Geological Influences

The climate requirements of *Rosa fortuita* are similar to those of *Rosa lignitum*, a rose species widely reported from fossils dating from the Oligocene to Miocene epochs in central Europe. Many plant species went extinct or were restricted to small regions in Europe during the dramatic climate changes that have taken place since the Pliocene epoch (5.33–2.58 million years ago.) However, the complex topography in southwestern China may have helped secure the survival of many plants, including roses, from climate change by allowing plant communities to migrate to regions with suitable environmental conditions.

The present-day diversity of roses in China coincides with the uplift of the Qinghai-Tibet (Himalayan) Plateau. The exact pattern and timing of this land shift are not understood, but the region's geology greatly affected the topography and monsoonal climate, and therefore contributed to species diversification. There are rugged mountains in southwestern China caused by the collision of India with the Eurasian continent, generating complex topography and climate in this region. This variety of habitats has provided opportunities for the diversification of roses. For example, there are currently seven species of roses on the Meili Snow Mountain, the highest peak in Yunnan Province, with a difference in elevation of approximately 15,748 feet (4,800 meters).

Climate Conditions and the Diversification of the Rose

Recent molecular studies reveal that the complicated topography and climate in southwestern China have played significant roles in the high diversification of many plant groups, including roses. The range of climate conditions in this area is associated with the uplifting of the Qinghai-Tibet Plateau since the late Miocene epoch, which has enabled the general climate requirements of roses in southwestern China to have remained the same for millions of years. It is fascinating to think about the rose's existence for millions of years despite all—or possibly because of—the geologic and climatic changes.

From millions of years ago to the beginning of ancient cultures, the rose's story remained silent. Simply, this quietness is because there was no one around to record its history. Subsequently, ancient societies bring the rose back into view and reveal the rose's importance to their cultures.

BELOW The specific epithet "fortuita" means it is a fortunate coincidence that this fossil was found. According to morphological comparisons, the fossils of *Rosa fortuita* most resemble *Rosa helenae*, seen here, a species found in southern China, Thailand, and Vietnam.

FIRST RECORDINGS OF THE ROSE WITHIN CULTURES

STORIES OF THE ROSE began to appear in various cultures around 3000 BCE. Confucius (551–479 BCE) writes of the rose's importance to Chinese culture. Within this history, the Zhou Dynasty (1046–256 BCE) was the longest-lasting lineage to date. Confucius writes that the emperors regarded roses highly and planted them in the Royal Gardens of China. The cultivation of roses was also widely present during the Han Dynasty (141–87 BCE). It was not until much later, in the eighteenth century, that Chinese roses (*Rosa chinensis*) were introduced to Western cultures and modern roses appeared.

The Legendary Rose of Persia

In northern Persia, the province of Faristan touts the birthplace of the cultivated rose. It later spread across Mesopotamia and eventually to Greece. Faristan was the center of production of rose water, exporting all over the world. The caliph, the chief Muslim civil and religious ruler in Baghdad, annually received 30,000 bottles of rose water from this region. Rose petals, oil, and essence were used as a perfume, in food, and for medicinal and religious purposes.

For more than 5,000 years, China and Persia were recorded as being the only places where naturally fragrant rose varieties (and yellow flowers) grew. The single-layered Iranian rose that grows in Qasmar, near Kashan, bears

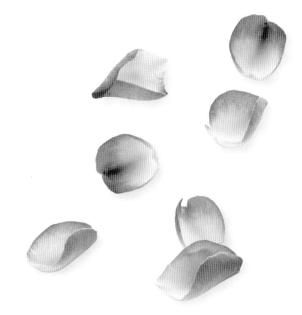

such an exquisite perfume it is grown solely for its oil. Modern tourists can visit to see the traditional method of extracting oil from the blossom. Such was the splendor of the rosewater fragrance—*gool ab* (gool = flower, ab = water)—that the oil itself, by its very name, became the rose.

THE ORIGINS OF ROSE WATER

Rose water and rose oil obtained by hydro-distillation of fresh rose material have a prolific presence throughout the centuries. According to some historians of the distillation process, there is evidence that people of the ancient Indus valley civilization in present-day India and Pakistan were distilling floral waters using earthenware ceramic pots in the fourth millennium BCE.

Archaeologists have discovered ceramic pots that may have been used in distillation. At around the same time, Assyrians and Sumerians in the Mesopotamian civilization may have been distilling floral waters.

Mesopotamian tablets and jugs reveal that the art of extracting perfume date from 3500 BCE. Boiling water is used to release these rose scents and aromas. The Assyrians became famous for this method of extraction.

The use of rose water in beauty rituals dates back to the seventh century BCE when the ancient Persians influenced and encouraged the use of the rose in India.

Persian Poetry and Roses

Persian poetry often references the rose. Omar Khayyam, the eleventh-century Persian poet and philosopher who wrote the famous *Rubaiyat*, was a passionate lover of the rose and told one of his pupils that his tomb "will be in a spot where the north wind may scatter roses over it."

A nightingale is an object of longing and adoration for the rose. Legends tell of how the red rose got its color. Initially, roses were all white, until the nightingale saw the rose and fell deeply in love. At this stage, nightingales merely croaked and chirped like any other bird, but from that point onward the nightingale's love was so intense that he was inspired to sing for the first time. Eventually, the nightingale pressed himself to the flower, and the thorns pierced his heart, giving the color red to the rose forever. Asian writers of this period also represent the nightingale as sighing for the love of the rose.

The Zoroastrian text *Bundehesh* from the eighth and ninth centuries mentions "hundred-petaled" roses plus a "dog" rose. It states that the rose produced thorns only when evil appeared in the world. This idea is that the rose's allure represents the enthusiasm for living and the thorns symbolize challenges.

Sa'di, a Persian poet, called his famous compilation of moral and religious poems from 1258 the *Gulistan*, or "rose garden," to imply the pleasure and insight they would bring to the reader.

Roses were said to be born of drops of Mohammed's sweat, according to a legend that was common in Turkey and Persia. This story was both a cause and effect of the growing popularity of the flower. Other legends say the rose replaced the lotus as the queen of flowers. In the Near East, roses came to symbolize faithfulness and everlasting affection, extending even after death. The rose has certainly claimed importance in Islam: in Muslim countries, the rose was considered so holy that a mosque desecrated by nonbelievers could only be cleansed by being washed entirely with rose water.

Roses were mentioned by the philosopher Vatsyayana, who wrote the *Kama Sutra* between the first and third centuries CE. He said one of the duties of a virtuous wife was to tend a garden, "the China Rose and others should likewise be planted together with the fragrant camel grass and the fragrant root of the vetiver plant" (*The Kama Sutra of Vatsyayana*).

LEFT A depiction of the 'Omar Khayyám' rose. A quote from Khayyám states: *"Be happy for this moment. This moment is your life."*

ASIA: ROSES AND RELIGION

THE FOUNDERS OF AYURVEDA, separately Charaka and Sushruta, are believed to have lived in the first century BCE, giving testimony of roses during that time. In the Charak Samhita (the ancient language of Sanskrit), the rose is classified in eight different ways. One example is "Saumyagandha," meaning having a pleasing smell. Each name was a classification of a particular characteristic of the rose's attributes, many of which were curative.

In the oldest sacred and spiritual works in Zend (Avesta), in the teaching of ancient Persia and Sanskrit, and in the superb records of ancient India, the rose always plays a symbolic role in the creation of the world and of humankind. Fable has it that the two gods, Brahma, the creator, and Vishnu, the protector or preserver of life, who resided in the Himalayas, had a debate about which was the most beautiful flower—Brahma promoted the lotus, and Vishnu the rose. After seeing Vishnu's celestial arbor loaded with fragrant roses, Brahma conceded the supremacy of the rose.

In Hinduism and Buddhism, the shape of the rose brings to mind the cup of life or the center of Mandala, a geometric arrangement of shapes said to symbolize a meditative path to Nirvana. Siddhartha Gautama the Buddha lived in the sixth century BCE. The Gandhara School of Sculpture (first to second century CE), which is an amalgamation of Indian and Greek traditions, depicted the Buddha in many statues, most of which show him seated on a pedestal of lotuses. However, a few portray him on a five-petaled rose. In ancient India, the cosmic rose, known as Tripura Sundari, symbolized the beauty, strength, and wisdom of the divine mother.

The Legendary Rose of India

Ancient history holds several pieces of evidence verifying the existence of the rose in India and pointing to it being very much a part of its social, medical, cultural, and religious makeup. According to accounts, Emperor Shah Jahan, the famed Mughal ruler of India, was symbolized with a red rose and left behind a grand legacy of structures constructed during his reign. His most significant building was the Taj Mahal in Agra, India. Many rose motifs adorn the walls here.

Legend has it that the Indian god Vishnu created his bride, Lakshmi, from rose petals: 108 large and 1,008 small. And so the rose became a symbol of beauty. Another legend recognizes wild roses as confirmed by the medical treatises of Charaka and Sushruta (the aforementioned founders of the Indian system of medicine called Ayurveda). Different rose species held unique medicinal properties, and each was allocated an original Sanskrit name that suggested its curative properties for specific ailments.

LEFT Rose decoration featured on the Taj Mahal is one example of the ornamentation that depicts forty-six different species of plants found there.

Roses and Trade

Trade between India and China took place as early as the second century BCE as a Yue Chi chief carried Buddhist scriptures to China (after Kasyapa Matanga introduced Buddhism to the country). Chinese visitors said it was customary to see rose garlands at Buddhist sites. Trade with China continued over the centuries by the different kingdoms ruling separate areas of India—rose products were especially popular. A stone pillar located at Motupalli, near Guntur, bears an inscription dated 1244 CE. It states how King Ganapati, of the Kakatiya Dynasty of Warangal, waived the customs duties on some items from China—this includes rose water. In 1300 CE, Rashid-ud-din, a Muslim traveler and chronicler who visited Gujarat state in western India, noted that "the people were very wealthy and happy and grew no less than 70 kinds of roses."

The Vijayanagara Empire was a great dynasty of south India from the fourteenth century to the sixteenth century. Accounts from many travelers revealed that roses were fundamental to the daily lives of both nobles and ordinary citizens. Abdur Razzak was a Muslim diplomat from Persia who visited this royal court in 1443. He wrote: "Roses are sold everywhere. These people could not live without roses, and they look upon these as quite as necessary as food."

Portuguese travelers Domingo Paes and Fernaz Nunis journeyed to India around 1537, and talked of the many plantations of roses they saw, bazaars where baskets packed full of roses were sold, and gardens of the nobility in which rose plants grew in abundance. Men and women alike used roses as adornment. The king would offer his morning prayers and shower white roses on his favorite followers, while his elephants and horses were all decorated with rose wreaths. The traveling duo described the pillars and walls of the king's bedchamber as, "at the top had roses carved out of ivory."

The Mughal emperors arrived to rule in India in the sixteenth century, bringing with them camel-loads of roses. They came from Persia and Afghanistan, and the first of them, Emperor Babur, is said to have introduced the first Damask rose. This royal dynasty was intrigued by the Mughal style of gardening, and Emperor Jahangir had the Shalimar gardens of Kashmir created for his empress Nur Jahan—they are well known to this day. The Mughal garden celebrates the use of water features, such as fountains, pools, canals, and running water.

RIGHT Nur Jahan, whose name means "light of the world," was the wife of the Mughal Emperor Jahangir. Her family is associated with the discovery of the attar of rose (see page 180).

British Trade with India

The British traveled to India initially as the trading company The East India Company in the seventeenth century. As their ships sailed from China carrying merchandise to England, they stopped for refueling at the port of Calcutta. Here, they traded items that included plants (such as roses), which were kept at the Botanical Garden in Howrah. Sir William Roxburgh started this garden in 1793, and each consignment of plants imported by the British would be planted here. One of these plants was 'Fortune's Double Yellow' rose; others included Chinas, Noisettes, and Early Teas.

Indian Rose Products

Rose products tend to be distinctively Indian and come in many forms: cosmetic, medicinal, and dietary. Roses form the base of these, and to make them, tons of rose petals are dried in the shade to produce *pankhuri* (meaning "flower petal"). This is then sent to the Middle East every day and is used in beverages, foods, and medicines. The cooling effect of roses is evident in numerous medications and unguents. The flavor of rose is added to foods and sweets and it is a pleasurable fragrance at occasions such as weddings. *Ark Prakash* is an old Sanskrit text that mentions rosewater distillation, and Nagarjuna, a famous Buddhist monk who lived in the eighth or ninth century CE, has provided details on how to distill rose water.

Ruh gulab is another name for the oil of roses. In Indian cuisine, we find rose oil in rose essence, rose syrup, rose sherbet, rose wine, rose liquor, rose honey, and rosehip jam.

Bulgaria, Morocco, France, and India are the dominant rose oil-producing countries. India uses *Rosa damascena* and *Rosa bourboniana* to make rose oil. It is made using steam distillation of rose flowers plucked very early in the morning. The process produces rose water from which, over days, rose oil is collected in minute quantities. It takes 4 tons (4,000 kg) of petals to produce half an ounce (10 grams) of rose attar.

LEFT 'Fortune's Double Yellow' rose can climb up to 40 feet (12 meters). It was discovered in 1844 by Robert Fortune in China. Characterized as covered with blousy copper-colored blooms, the actual colors are a vibrant apricot-yellow with a hint of rose or crimson on the outside petals.

THE DELIGHTS OF TURKEY

GEOGRAPHICALLY AND CLIMATICALLY, **roses have flourished in the area known historically as Mesopotamia for many millennia. It seems likely that they endured and were used there in ancient times, especially since we have learned from archaeological excavations that they were known and used in the first millennium BCE in the Peloponnese, a peninsula in southern Greece.**

Twenty-four species of rose are recognized as growing in Turkey and the East Aegean Islands. Cuneiform tablets reported roses some 5,000 years ago, and Assyrian tablets told of roses and rose water. We are unable to identify the rose species mentioned in these ancient texts, but as its scent was commended, it suggests fragrant rose species such as *Rosa gallica* or *Rosa damascena* of Anatolia. Cuneiform tablets reveal the process of boiling roses for their oil. The tiny quantities again illustrate how precious it was.

Ancient Mediterranean Pleasure Gardens

Mediterranean civilization highlights the use of plants and flowers for ornamentation. The first gardens were seen at the eastern end of the basin and the idea of the pleasure garden was taken to the west as a result of trade and colonization. It developed from the widespread fruit, flower, and herb cultivation. This is, to a greater or lesser degree, dependent upon summer irrigation. The use of walls for enclosure gave protection to valuable fruits, orchards, and vineyards. Trees were set out in rows with flowers often planted in the spaces between them—this meant a more economic use of the precious soil. These flowers were both useful and beautiful, and included species such as the saffron-yielding crocus, the edible poppy, or the violets, irises, and roses of Boeotia (a region in central Greece), Cyrenaica (the eastern coastal region of Libya), and other lands in order to produce perfumes and ointments.

THE ROSE ON CUNEIFORM TABLETS

The "plant lists" and the tablets called "medical texts" have revealed much about the kinds of plants raised and how they were used in pharmacology in ancient Mesopotamia. Some of them have been easy to identify, but the majority are more challenging to link to known flora.

One scholar, Reginald Campbell Thompson (1876–1941), was satisfied that the word "Kasu" meant rose. It occurs 181 times on the 600 excavated tablets. "Kasu" was later identified to be the word for "Dodder," which flavors beer. The term "amurdinnu," which is in some of the medical texts, may have referred to a "bramble" or "wild rose," which Thompson somewhat grudgingly agreed.

Igor M. Diakonoff, a Russian scholar, cites its use in the *Epic of Gilgamesh*. In this poem from ancient Mesopotamia, "amurdinnu" appears to be a thorny flower with a strong smell (which, as Diakonoff points out, a bramble does not have, but a wild rose does).

ANCIENT EGYPTIAN GARDENS

DURING THE GRECO-ROMAN PERIOD, mummies in Egypt continued to be fitted with floral decorations. The application of petals in wreaths included a wide variety of plants essential to the Egyptians, including roses. Theophrastus, in his writings, described roses in considerable detail, illustrating methods of propagation and their blossoming periods in Egypt.

Flowers were used extensively in the social and religious lives of the early Egyptians. At banquets, they embellished both tables and guests. Flowers worn by guests and held in their hands were a subject of adoration from other guests. Garlands festooned the wine jars; bouquets were presented to the gods; wreaths encircled the necks of sacrificial geese and gulls. Egyptians also believed that the rose had healing and aphrodisiac properties. Households would often boil roses down into an oily residue, which they would then use as a beauty-enhancing cosmetic balm. Taken together, these attributes emphasize the rose's importance in Egyptian culture, including the use of them in burial rites, in tombs, and in mummified bodies.

Egyptians used roses symbolically as a connection to the afterlife. They regarded them as a metaphor for being born in the spring and living in the sun. During this time—and very much like the present—the real legacy of roses is that they were a symbol of love.

ANCIENT FUNERAL WREATHS

Funeral wreaths dating back to 170 CE are the earliest-known record of existing roses. Substantially preserved wreaths found blanketed with dust and sand were scarcely changed. These wreaths provide the closest means for examination and comparison with existing plants. The roses had been picked in an unopened state, to prevent the petals from falling. In drying in the coffin, the petals had shriveled and shrunk into a ball, and when moistened in warm water and opened, the set of stamens rises in a beautiful state of preservation. After analysis of specimens of the wreaths at the Royal Botanic Gardens, Kew in London, England (where several are still held in the Herbarium), recognized roses are said to be the Holy Rose or *Rosa sancta*, also known as *Rosa richardii*. This was the rose most used in funeral wreaths and to adorn burial chambers. The other rose—*Rosa gallica*—still survives to this day and is cultivated throughout Europe.

RIGHT Depiction of an Egyptian Pharaoh lying in state, bedecked with wreaths and other floral offerings.

EVIDENCE OF ROSES WITHIN THE ANCIENT CULTURES

EMMETT L. BENNETT JR. (1918–2011), the scholar who decoded the symbols used in Linear B, a syllabic script used more than 3,000 years before the Greek alphabet was formed, was asked to examine the Olive Oil Tablets of Pylos. He found they gave exceptional detail about the quantities of and transactions conducted with oil. Bennett's analysis of the symbols on the tablets led him to conclude the principal commodity described is oil—likely olive oil—particularly that used for anointment or perfume.

The secondary oils used were rose-scented, sage-scented, and cypress-scented. The use of olive oil in antiquity for the production of aromatic oils and unguents is relatively well attested. Most references to olive oil in Linear B tablets from both Pylos and Knossos are related to its use as an essential ingredient for the manufacture of aromatic oils and ointments.

In the eighteenth century, there were claims that some very ancient coins dating back to 3000 BCE and bearing "some impression of a full-blown rose" (Phillips and Rix) were in the graves of the Aryan people called the Tschudes. No one knows if they ever existed.

RIGHT *Helios on the Sun Chariot* (c. 1777) by Andreas Brugger is a depiction of Helios's chariot drawn by four steeds. In Greek mythology, Helios was the god of the sun. His riding of the golden chariot led the sun across the sky each day from the east (Ethiopia) to the west (Hesperides).

ABOVE The two sides of an early coin reveal the face of Helios, the Greek sun god, and a rose.

The oldest coins with roses imprinted on them date back to c. 500 BCE and come from Rhodes. Archaeologists believe the island was named after the nymph Rhode of Greek mythology, whose symbol was the rose and who was loved by the sun god Helios. Rhodes was the center of the cult of Helios, which lasted from 400–80 BCE. His head is displayed on one side of coins found in excavations there. A rose is represented on the other side.

If Zeus chose us a King
   of the flowers in his mirth,
He would call to the rose,
   and would royally crown it;
For the rose, ho, the rose!
   Is the grace of the earth,
Is the light of the plants
   that are growing upon it!

Song of the Rose, attributed to Sappho,
version by ELIZABETH BARRETT BROWNING, 1893

Rosa centifolia

3

THE STORY OF THE GREEK ROSE

From the ancientness of the rose some 35 million years ago, the rose's narrative continues with the Greeks. A reality, authority, and a physical presence of the rose can be found in Greek gardens, in art and pottery, and in writings and storytelling. Thus begins the rose's influence in many aspects of life among the Greeks: botanical, medicinal, pleasure, ceremonial, and above all, symbolic.

THE DEVELOPMENT OF GREEK GARDENS

WITHIN THE GREEK WAY OF LIVING, we start to learn the notion of gardens created for the idea of pleasure. Instead of a routine, utilitarian, yet vital plot of land used for growing crops, the new idea of parcels of land used for enjoyment, exercise, and beauty begins to take shape in a naturalistic way.

Going back to Homeric times (1100–800 BCE), the Phoenicians were a civilization of city-states that created an abundant life along the Mediterranean. Their sophistication and organization allowed for the trade of plants and trees all across the Mediterranean. Trade routes made their way from Persia to Egypt, and eventually to Greece and Rome. In *The Story of Gardening* Penelope Hobhouse comments: "Greece has more than 6,000 species of flowering plants and trees." This abundance laid the groundwork for the future of gardening as we know it,

even today. Trade with Egypt brought roses, perhaps from the Nile region. *Rosa richardii*, *Rosa canina*, and *Rosa centifolia* were thought to come from Macedonia. Around the fifth and fourth centuries BCE, roses were grown extensively on the island of Rhodes. It seems this influx of plants into Greece naturally led to the conception of adding plants to the public garden space for the pure pleasure of beauty. Pleasure gardens are common areas for the intention of community gathering in a place of beauty, leisure, exercise, and amusement.

Planted Gardens

Decorations of the pottery of the palaces give evidence of Greek gardens and their joy of plants. In their book *Garden Lore of Ancient Athens*, Thompson and Griswold cite that references to such fantastic landscapes did indeed inspire the potters of the time to delineate these botanicals on their pieces of pottery. This pottery illustrates plants and flowers in its features and embellishments. It is interesting to note that despite these artistic artifacts of pottery and paintings displaying horticultural references, the existence of vast books written on gardening techniques is hard to find. Any reference in books would support the idea of these pleasure gardens. However, we can see evidence of more highly developed gardening skills among the gardens of the temples, fountains, and parks themselves.

Greek land developed as gardens started to include flowers grown for wreaths, altar decorations, and fragrance. Flowers supplied the fragrance for perfumes and ointments (both medicinal and cosmetic) present in daily life. Some regions became known for producing exceptional plants. For example, Cyrenaica (part of Libya then known as Kyrenaika), a later Greek colony in the seventh century CE, was particularly famous for its roses, violets, and crocuses. These flowers were described by Ellen Semple in "Ancient Mediterranean Pleasure Gardens" as being "more fragrant than elsewhere" due to the growing conditions produced by the semi-arid climate.

LEFT Flowers and botanicals depicted on Greek pottery show the importance plants had in the Greeks' daily lives.

Elizabeth MacDougall states in *Medieval Gardens: History of Landscape Architecture Colloquium* that in the early Greek Mycenaean cultures (1600–1100 BCE), according to archaeologists, the palaces of the time did not have planted gardens. The plantings came in pots that were placed in courtyards. Penelope Hobhouse's *The Story of Gardening* confirms that the Minoans grew plants in containers. These included plantings of pomegranates, myrtles, lilies, irises, and roses. The difference between a few pots and a full garden gives a clear distinction between the newness of Greek gardens and the much later Hellenistic period gardens that were influenced by a later, more developed Greek culture (Hellenistic refers to a time after Alexander the Great in 323 BCE and the beginnings of the Roman Empire). Through the Hellenistic lens, the gardens became grand, walled paradises. It is in this later time in Greek culture that the rose is very present.

LEFT Early Greek gardeners grew plants in pots that were important to the Greek lifestyle, such as myrtles, pomegranates, and roses. It wasn't until later during the Hellenistic times that these gardens became magnificent, walled paradises.

Interior Gardens

If we look at the structure of the Hellenistic home it often had an open court with interior gardens surrounded by columns. In the Roman ruins of Pompeii and Herculaneum, this peristyle method of architecture is easily seen (peristyle refers to a row of columns that usually surrounds a courtyard or building). This row of columns, or colonnade, allows for a covered walkway and provides shade from the sun and protection from the weather. The interior courtyard, which often included statues, was an accepted addition to the home. The garden beds in these courtyards often included lilies, violets, and roses. Trees and fountains were cooling during the warmer months. Great skill is evident in the Greeks' ability to manipulate water for the use of irrigation and fountains.

Parks, Temples, and Cities

In classical Greece, we see further gardening skills in the landscapes of the public parks devoted to exercise and recreation. The Academy of Athens is such a place embellished with fountains and walks. Plato established and taught at this college on the outskirts of Athens around 390–380 BCE, and it persisted throughout the Hellenistic period. Trees planted were so abundant that they provided immense shade from the afternoon heat. Groves of laurels and olives, as well as narcissuses, crocuses, and roses grew in profusion.

Copses of trees and other plants adorned the temples of the time. These include the shrines devoted to Aphrodite, Artemis, and the nymphs at Olympia. Furthermore, accounts of the garden of Daphne, sacred to Apollo near the city of Antioch, rate it as one of the finest. William Smith's *Dictionary of Greek and Roman Geography* gives detailed accounts of these temples. Temple gods and goddesses were often associated with particular flowers, for example, violets, myrtle, and roses were all linked with Aphrodite. This floral

LEFT This painting from the School of Fontainebleau, France, shows Venus (Aphrodite) at her morning toilette surrounded by bays, myrtles, and roses.

GREEK FLOWERS

In the Greek lifestyle, the theme of flowers becomes even more present. Theophrastus (c. 350–287 BCE) writes in *Historia Plantarum* of plants cultivated to make wreaths and garlands—these include roses. Greek celebrations and ceremonies made the insistence for flowers constant all year long. This demand was satisfied with violets in winter, roses in summer, and the colors of plants from fall.

representation for each god gives reason for the addition of flower gardens into the revered groves of trees around the temples. Aphrodite's association with flowers is highly illustrative in Greek mythology. These "sacred" groves of trees were eventually transformed and diversified with many ornamental plants. The fruit trees continued a presence. The flowery groves surrounding the temples are reported by Strabo, a Greek historian and geographer, and further reports come from Pindar, a Greek lyric poet, of a "pleasant garden" surrounding Aphrodite's temple.

Cities began to demonstrate a personality that included pleasure gardens and botanicals. Seleucus I Nicator (end of the fourth century BCE) founded the city of Antioch. The city's exceptional size and scale of design allowed the avenues to be planted in flower gardens, surrounded by colonnades, marble pavilions, baths, and fountains. An avenue of vineyards, rose gardens, and groves led through a suburb to the Park of Daphne and villas along this avenue adorned with their very own garden motifs added to the scenery.

A violet-wreathed city

Athens itself is botanically described as a "violet-wreathed" city. There is an astonishing demand for violets, among many other flowers of the time. These ornamentals are ever-present in the wreaths and garlands worn by Greeks during festivals as well as everyday life. Athens possessed flower markets where many varieties of flowers were sold, including roses. In such a large city of structures, these markets were necessary as interior garden space in the city was limited. The lack of space meant gardens in the "suburbs" helped meet the excessive demand for the flowers. Roses were particularly welcome, and were often transplanted and cut back to encourage new growth and continual blooming. These suburbs became a "green belt" around the city.

Rose Wreaths

The wearing of rose wreaths had symbolic connotations. Solon (c. 640–560 BCE) was an Athenian statesman, lawmaker, and poet. In order to uphold his self-imposed morals on the city, he sponsored the decree of who could wear rose wreaths. Non-virgins were prohibited; wreaths were presumably reserved only for virgins, symbolizing their innocence and purity. At about the same time, Stesichorus (c. 602–555 BCE), a Greek poet who lived in Himera, a town in northern Sicily colonized by the Greeks, wrote that the rose was associated with drinking and feasting. In Stesichorus's heroic ballads, he tells of rose wreaths that revelers wore at banquets. Rose-wreath symbolism continues with Eros, son of Aphrodite, and god of love and sexual desire. He often flaunted style by wearing a wreath of roses, as does Dionysus, the god of wine. This association between roses and revelry was later to reach its peak in Roman times.

With increasing trade and imports, the Greeks were often looking for the latest species of flowers and plants to add to their garlands and wreaths. They knew which countries were growing the best and most fragrant varieties of different flowers. The propagation of these new species required considerable skill to get these plants to survive in their non-native environments and soils. In Greece's ancient classical kingdom of Macedonia, the city of Philippi had a hundred-petaled rose that had been propagated from a variety that grew wild near Mt. Pangaeos (Mt. Pangaion). Ellen Semple quotes the legend that says that mysterious rose might have been from the garden of Midas, "in which wild roses grow, each one having sixty petals, and surpassing all others in fragrance." References to this King Midas rose are numerous. It is sometimes described as *Rosa damascena semperflorens*—aptly referred to as "the King's rose." Today, it bears the name 'Autumn Damask.' This is the only old rose that has a second bloom in the fall. It is well proven that the later cross of the China genetics with the European roses made the modern repeat-flowering roses possible.

GREEK WRITINGS AND THE ROSE

HOMER'S VERSION OF THE PLEASURE GARDEN reveals itself in *The Garden of Alcinous,* where he writes of a mythical garden with "unearthly lushness." This garden, though planted for practical purposes, reveals a sense of horticultural beauty. From the translation by Robert Fagles, Homer describes it as follows:

Here luxuriant trees are always in their prime. Pomegranates and pears, and apples glowing red, succulent figs and olives swelling sleek and dark. And the yield of all these trees will never flag or die, Neither in winter nor in summer, a harvest all year round.

The Odyssey, HOMER, translated by Fagles, 1997

The Greeks cultivated the rose during the time of Homer. In *The Iliad* and *The Odyssey,* both written in the eighth century BCE, Homer wrote about the rose metaphorically. He compared the beauty of a rose to the colors of the sunrise using the mythological Aurora as she "flies across the sky" each morning with the colors of dawn: "Here she is again, with the roses!" Homer describes Aurora with fingers of roses, and she perfumes the air with them as she travels across the heavens.

The ancient Greek god Harpocrates also represented the newly discovered sun. The Greeks adopted his name from the Egyptian god Horus, who was a

symbol of the sun emerging into daybreak. Harpocrates was also known as the god of silence and secrecy. Thus stemmed the expression "sub rosa," stating that all said was to remain a secret (in Latin, the phrase *sub rosa* means "under the rose"). It was custom to suspend a rose over the table in the dining area as a reminder that expressions stated within the room should remain silent elsewhere.

Theophrastus

Theophrastus (371–287 BCE) was a native of the Greek island of Lesbos. He is often considered the "father of botany" for his work with plants. His interests also included philosophy and science, and he was the first primary botanical writer.

Theophrastus described roses in considerable detail, explaining methods of propagation and their blossoming

period in Egypt. The extended growing season and fertile soils of Egypt give some possible clues to how the sheer quantity of roses demanded by the Greek lifestyle was satisfied.

Epicurus

Epicurus (341–270 BCE), a Greek philosopher, had his private rose garden just outside Athens. Here, Epicurus established his school, known as "The Garden." Some writers say that his roses helped to achieve his high degree of tranquility. Epicurus argued for a "retreat from chaos and politics and to stay in one's countryside and tend the roses." Although this soothing idea sounds pleasing, most Greek towns were too small to allow for any rose garden that wealthy citizens like Epicurus enjoyed. Most residents who did have space to garden grew roses for ornamentation of their wreaths and used the oil in their homes to keep illness at bay.

Herodotus

The Greek historian Herodotus (c. 480–425 BCE), known as the "father of history," writes of roses that had sixty petals. They were grown in the garden of King Midas in Macedonia and are described by Herodotus as having the most potent perfume of any other rose. He identifies this sixty-petaled rose to be *Rosa gallica* or *Rosa damascena*, or possibly even a double form of *Rosa alba*. Perhaps inside these magnificent Macedonian gardens is where Alexander the Great (see page 70) first became infatuated with the rose.

Sappho

The Greek poet Sappho (630–570 BCE), also from Lesbos, famously penned verses to be arranged to the music of a lyre and sung. In her time, Sappho was considered one of the greatest poets and often wrote about love, desire, and jealousy. In her sweet words, she writes about the tranquility of the garden (see right). This garden sounds pleasurable indeed.

Very little of Sappho's poetry still survives, and just one poem, the *Ode to Aphrodite*, written around the sixth century BCE, is complete. In this poem, the speaker calls on the help of Aphrodite in the pursuit of a beloved. Sappho wrote: "The rose each ravished sense beguiled" and called the rose the "queen of flowers"—images that would be used repeatedly over the centuries. Sappho established roses in the temple of Aphrodite, summoning the goddess to rise in her "graceful grove of apple trees" amid "altars smoking with frankincense."

RIGHT *Sappho and Erinna in a Garden at Mytilene* (1864) by Simeon Solomon depicts Sappho embracing a fellow poet, Erinna, who belonged to a group of women devoted to Aphrodite and the Muses.

> *Through orchard-plots with fragrance*
>   *crowned*
> *The clear cool fountain murmuring flows;*
> *And forest trees with rustling sound*
> *Invite to soft repose*
>
> SAPPHO

Anacreon

Another Greek lyric poet, Anacreon (570–475 BCE), composed an ode committed to celebrating all aspects of the rose's beauty: its perfume, its power to heal, and the esteem with which it was held among the gods. Anacreon's poem *The Rose*, which includes the extract (see right), is an excellent summation.

Rose, the gods' and men's sweet flower;
Rose, the Graces' paramour:
This of Muses the delight,
This is Venus' favourite;
Sweet, when guarded by sharp thorns;
Sweet, when it soft hands adorns

ANACREON

ALEXANDER, THE GREAT ADMIRER OF ROSES

ALEXANDER THE GREAT is widely known for his military prowess and for establishing the largest empire of the ancient world. Less well known is his love of plants. During his invasions of Persia he collected plants (including roses). Plants were sent back to Aristotle and Theophrastus from as far off as India.

Perhaps it was his learning under Aristotle until the age of sixteen that gave Alexander the Great a knowledge of and passion for botany and roses. He is given credit for introducing roses to the rest of Europe and Egypt.

Flowers feature heavily in Egyptian wall paintings and hieroglyphs from as early as 2500 BCE and the rose was a common motif. The earliest-known record of a rose is an Egyptian hieroglyph inside the tomb of Pharaoh Thutmose IV. Although roses were associated with the Egyptian goddess Isis, it was Alexander and his Greek influence that led the Egyptians to wrap the rose into their culture.

LEFT Alexander the Great (356–323 BCE) was highly influenced by Persian culture. Here he is depicted entertaining the Mongol Ambassador on mats in a rich floral landscape.

ALEXANDRIA

The port city of Alexandria in Egypt, near the mouth of the Nile, is just one of the cities to carry Alexander's name after it was conquered. He arrived in Egypt in 332 BCE and having only stayed there a short time, he was able to implement vast Greek influence. One of these influences brought the rose into the mainstream of Egypt. Later, Cleopatra, a descendant of Ptolemy I Soter, one of Alexander's generals and co-ruler of the Ptolemaic dynasty (305–30 BCE), which was the last great dynasty of the Hellenistic period, saw an escalation of the rose's fashionableness.

THE STORY OF ROSES IN KNOSSOS

THE HISTORY OF KNOSSOS ENDURES from ancient Greek references to this major city on the island of Crete. Knossos was known to be inhabited in the Neolithic period, sometime in the seventh century BCE. Shortly after 1700 BCE, the palace and the city of Knossos boasted a population of 100,000 people. Cretan resources, such as oil, wine, and wool, created prosperity and allowed Knossos to develop its trade and influence in the region.

Herodotus tells of the legendary King Minos of Knossos and his creation of a "sea empire." The Minoan civilization flourished from about 2700 BCE to around 1450 BCE, and dominated the Aegean with its trade and culture. It was the earliest civilization of its kind in Europe. Archaeological evidence of Minoan palace pottery supports the idea of its expansive influence and achievements. Discoveries include pots that contained oils and ointments at sites throughout the Aegean islands, as well as mainland Greece, Cyprus, and along coastal Syria and Egypt.

The history of Knossos shows that it had been destroyed several times by earthquakes, invasions, and volcanoes; but rebuilt to more elaborate and complex versions of itself. A local amateur archaeologist, Minos Kalokairinos, discovered the ruins of Knossos in 1878 on land owned by his father. Through his efforts, we can now understand the importance of Knossos to Minoan civilization and culture. Excavation revealed elaborate labyrinths, floor plans of the palace, a compact city, elaborate pottery, and houses with wall paintings.

The Earliest Known Rose Painting

Further archaeological excavations at Knossos confirmed that the rose was a part of Greek life. The English archaeologist Sir Arthur Evans (1851–1941) discovered the world's earliest known painting of a rose on the *Blue Bird* fresco at the House of the Frescoes, dating back to c. 1550 BCE. He described the fresco as follows:

To the left, for the first time in Ancient Art, appears a wild rose bush, partly against a deep red and partly against a white background, and other coiling sprays of the same plant hang down from a rock-work arch above. The flowers are of a golden rose color with orange centres dotted with deep red. The artist has given the flowers six petals instead of five, and has reduced the leaves to groups of three like those of a strawberry.

The Palace of Minos, SIR ARTHUR EVANS, 1902

BELOW The *Blue Bird* fresco (c. 1550 BCE) from the House of the Frescoes found during excavations at the Palace of Minos, Crete. Components of the composition include a bird, irises, lilies, and roses.

Historically found in Crete, *Rosa pulverulenta*, a species
rose that has a pink bloom with a lighter center and a mild
pine fragrance. It fits the description of a rose that would
have existed at the time and location of the Palace of
Minos. It is also known as *Rosa glutinosa*.

A Problem with Petals

Sir Arthur Evans had begun excavations at Knossos in 1900, which continued for 35 years. Sir Arthur restored large parts of the palace in a way that it is possible to appreciate the grandeur and complexity of the structure that evolved over the millennia. Knossos is the largest Bronze Age archaeological site on Crete.

One might wonder who the artist of the *Blue Bird* fresco was and where the reference came from. We have learned that ancient palaces probably did not have gardens. Hence, it is thought that this rose might have been growing in a pot. According to *Isis, Rose of the World*, scholars believe Knossos did have roses and they may have been brought to Crete through trade with Syria.

The botanist C. C. Hurst (1870–1947) worked with Sir Arthur in England. Hurst identified the flower as a five-petaled rose and thought it bore "a striking resemblance to the Holy Rose of Abyssinia, Egypt, and Asia Minor." We now identify that rose as *Rosa sancta* (the modern-day rose is called *Rosa richardii*). Hurst's widow Rona, an archaeologist and a botanist, reopened the discussion of what kind of rose it was after she visited Crete in 1964. She supplied archaeological evidence that supported a different explanation to that of her husband's identification, and she described roses containing six petals.

The "petal number problem" was further complicated when the *Blue Bird* fresco was restored by a Swiss art expert, Émile Gilliéron (1850–1924). Under Sir Arthur's guidance, Gilliéron suggested that there were several roses on the fresco: some with six petals that have a yellowish tinge, and one, much fainter, pink rose, which appears to have only five petals with strong veining on them. According to Rona, this veining was a distinctive characteristic of Gallica roses.

A fourth opinion comes from the work on Minoan frescoes by Mark Cameron (1974). He says the six-petaled flowers occur with and without brown-veined leaves. Cameron identified these flowers as the wild dog rose, *Rosa canina*.

The sources above agree that the fresco contains the earliest-known painting of a rose. However, there are several ideas as to whether it is *Rosa sancta*, *Rosa gallica*, or *Rosa canina*. Jennifer Potter, in her book *The Rose*, identified it as *Rosa pulverulenta*, a species related to *Rosa rubiginosa*.

THE IMPORTANCE OF ROSE OIL TO THE GREEKS

THE PROMINENCE OF ROSE OIL in Greek culture suggests that roses were cultivated specifically for this purpose. In Grecian times, rose oil was not extracted by distillation—this came later. Rose oil was made by immersing rose petals in oils such as almond, sesame, and olive. The process required enormous quantities of petals.

An early reference to the use of rose oil for medicinal and funerary purposes comes in Book XXIII of *The Iliad*, which tells how the body of Hector, killed by Achilles to avenge Patroclus's death, was anointed with rose oil and then embalmed by Aphrodite. The goddess herself by day and by night, "washed the skin with rose oil."

Theophrastus writes about the varying number of petals of different roses, the roughness of their stems, their color, and their scent. He writes that they have five, twelve, twenty, or more petals, and that those with the sweetest scent come from Cyrene (a city founded in 631 BCE in eastern Libya, an ancient Greek colony) and are used for making perfume. Theophrastus also describes methods of propagation and their blossoming periods in Egypt, thus explaining from where significant numbers of roses came.

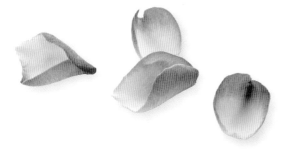

VARIETIES FOR ROSE OIL

There often seems to be speculation as to which rose variety would have been grown for the purpose of making rose oil. Pliny the Elder (23/24–79 CE), a Roman author and naturalist philosopher, wrote the encyclopedic *Naturalis Historia*. In their writings both Pliny and Theophrastus mention that the Greek rose was a "hundred-petaled" rose, suggesting that *Rosa centifolia* was a Greek rose. *Rosa centifolia*, often known as "cabbage rose," was cultivated later by Dutch rose breeders in the period between the seventeenth and nineteenth centuries. The earliest confirmed date of *Rosa centifolia* is c. 1318, negating Pliny the Elder's and Theophrastus's mentions.

LEFT Inspired by the palm tree's columnar shape, alabastrons (carved from alabaster) were used in the ancient world to hold oil, especially perfume or massage oils. This flask is attributed to the Painter of Palermo 1162 (c. 480 BCE).

EXPLORING ROSES IN ANCIENT TEXTS

OTHER WRITERS who mention roses include Ibycus, an ancient Greek lyric poet, who wrote praises to Euryalus, saying that Aphrodite nursed him "among rose blossoms." The tale of Chloris and Zephyrus states that she transformed the lifeless body of a woodland nymph into a flower. Her call for help was answered by Aphrodite, who added beauty to the flower, bequeathed the name "rose," and dedicated it to her son Eros, the god of love. Dionysus added nectar of fragrance, and the three Graces bestowed charm, joy, and splendor. Iris borrowed the rose's color, and Aurora painted the sky at dawn with it.

Eros

Another story mentions Eros. This god used roses as a bribe to Harpocrates (god of silence), seeking help to silence his mother's recklessness. As a result, roses would become conjoined with secrecy. An additional tale illustrates Eros stopping before a rose. While bending over to admire the beauty and smell the fragrance of the flower, a bee gathering nectar from it stung him. Eros quickly went back to his mother, Aphrodite, to tell of his distress. She gave him a "magical quill" to hurt the bee and settle the score. Eros then releases his quills upon the rose bushes. The rose's prickles represent his missed attempts.

ABOVE *Love and the Maiden* (1877) by
the Pre-Raphaelite artist John Roddam
Spencer Stanhope shows the myth of
Cupid and Psyche. These gods symbolize
everlasting love.

In the Greek myth of Cupid's marriage to Psyche, the god Jupiter was so pleased with this union that he instructed to make everything "glow with roses." This story is said to explain the proliferation or scattering of roses across the lands. Cupid, perhaps in his mischievous winged state, is said to have struck a bowl of wine, which fell to the floor near the god Dionysus. A rose bush grew from the spilled wine, which was then declared as a symbol of beauty to Aphrodite.

RHODANTHE AND
THE ROSE-TREE QUEEN

"The Rose-Tree Queen," from *The Wonder Garden: Nature Myths and Tales from all the World Over*, provides a myth about the creation of rose bushes involving Rhodanthe, the queen of Corinth, whose name means "rose bloom." Her beauty was such that she had three suitors who pestered her until she took refuge in the temple of Artemis, the virgin goddess of the hunt. Rhodanthe's beauty so struck visitors to the temple that they began to worship her instead of Artemis. Apollo, Artemis's twin brother, was so furious that he turned Rhodanthe into a rose bush. The punishment came to her would be suitors is described below:

*Her body was changed into a stem, and her head
became a large blushing Rose [...]*

*Her three lovers became a worm, a bee,
and a butterfly.*

FRANCES JENKINS OLCOTT, 1919

Midas

In addition to his garden being filled with roses famous for their superior fragrance, the well-known tale of King Midas's "golden touch" involves roses. The story tells of Midas's kindness toward Silenus, Dionysus's companion. Silenus was in a drunken state, and Midas offered Silenus repose to help him recover. Dionysus offered a reward to Midas for his kindness. Midas asked him for the gift of the golden touch— that whatever he might touch should be turned into gold. Caught up in the excitement of his new power, Midas touched an oak twig and a stone, and both turned to gold. As soon as he returned home, Midas touched every rose in his rose garden, and all became gold. In a later version of this story by Nathaniel Hawthorne, Midas's daughter came to tell him how upset she was that the roses had no more fragrance and had turned hard. Midas reached out to console her, but upon his touch, she also turned to gold.

Persephone and her nymphs were gathering blossoms when Hades, the god of the underworld, came to abduct her. Her bouquet included crocuses, violets, irises, lilies, larkspur, and roses. Modern-day florists use these same flowers, aware of their symbolic meanings, to compose a Greek-themed bridal bouquet aptly named Persephone's bouquet.

BELOW *In the Garden of Proserpina* (1893) by Harry A. Payne depicts Persephone, the daughter of Demeter and Zeus, who is the beautiful goddess of spring.

A Love Story

The rose was closely associated with
Aphrodite, the goddess associated
with love, beauty, pleasure, passion,
and procreation. Aphrodite was
symbolically represented by myrtles,
doves, sparrows, swans, and of course
roses, and the list of Aphrodite's
lovers is long.

Myths about Aphrodite capture
the dual nature of love that the rose
symbolizes: purity and innocence
represented by white roses; and sexual
passion and desire by red. In one story,
Aphrodite, as she quickly goes to care
for her wounded lover, Adonis, grazes
her skin on the prickles of a white
rose bush. Her blood settling on the
bush turns the roses red. White roses
become red, symbolizing how innocence
and purity change to fertility and
motherhood (see below).

The red rose
symbolizes fertility
and motherhood.

For every drop of blood that fell from Adonis's side, she
shed a tear like a pearl. [...] And the drops of blood grew
up into glowing red, red Roses! So, says the old Greek
wonder tale, this is how Anemones and Roses come
into the world.

FRANCES JENKINS OLCOTT, 1919

THE CEREMONIAL ADONIA

The ceremonial Adonia was to be held in late spring/early summer and observed annually by women in ancient Greece to grieve the death of Adonis. Flowers and plants are specifically grown to adorn the festival.

Theophrastus writes that roses were used in some pots to add to the plants grown for the event. The containers were mostly filled with lettuce and fennel seeds that grew quickly but then withered and died in the heat—a fitting symbol for a funeral.

Throughout the festival, the women of Athens danced, sang, and ritually mourned on the roofs of their houses. After the rooftop celebrations, the women picked the plants from the pots and descended to the streets. With these withered "gardens" in hand, they formed a small funeral procession before ritually burying the remains of the gardens at sea.

Misery and joy have the same shape in this world: You may call the rose an open heart or a broken heart.

DARD (1720–1785)

Greek mythology and storytelling abundantly show the rose's recurring and symbolic roles. The rose is a symbol of beauty, which also represents joy, beginnings, tragedies, and deaths. It represents the breadth of emotions: red roses of passion, white roses of purity, blushing roses of flattery—all are symbolic. The Sufi poet Dard observed (see left):

In the driest whitest stretch
of pain's infinite desert
I lost my sanity and
found this rose.

Rumi, c. 1260

Rosa gallica purpurea velutina parva

4

THE ROMAN ROSE OBSESSION

As we turn to Rome, we discover how the Greek way of life
influenced the Romans. As a result, the Greeks' love and use
of roses penetrated the Roman lifestyle from around the
time of the Greek settlements of southern Italy and Sicily
in the eighth century BCE.

What was a love of roses for the Greeks escalated into a passion
for the Romans, as rose use turned into rose infatuation.
Everything the Greeks had done with roses, the Romans did
more prominently and significantly. They wore rose wreaths;
added roses into their cuisine; used roses in a variety of cosmetics,
ointments, oils, and medicines; lay on rose cushions; painted
roses in frescoes; wrote of roses in myths; scented water
with roses; smothered their guests to death with roses;
and buried their dead with roses.

ROMAN CULTIVATION OF ROSES

THE ROMANS CELEBRATED the rose in art, gardens, banquets, rituals, architecture, and the daily lives of their citizens. This magnificent obsession with roses came with a high demand for their supply, which raises two questions: Where did all of these roses originate, and how were they grown?

Records show that roses were grown in the Middle East and Egypt during this time. These regions already had well-developed cultures, and the presence of the rose has previously been cited. Further, we know that from Greek and Hellenistic times roses had reached Europe.

The Romans indeed developed their gardens, and much writing can be found on how to cultivate them. In contrast to the Greeks, information about roses became widespread, including proper propagation methods and suggestions on cultivating roses commercially. The growing of roses in private gardens, and on roofs and balconies, became popular. Evidence can be seen in the preserved frescoes and the peristyle gardens of Pompeii (which we explore later). Individual rose growing allowed for a constant supply of roses for wreaths and garlands. This pleasant pursuit combated the smells and ailments of the city, and roses were also in demand for culinary and cosmetic purposes.

Roman Writings on Growing Roses

Instructions on how to grow roses are found in many texts by the scholars of the day. Perhaps such recognition was given because of the many Greek writers who laid the foundation for the Romans to take roses to the next level.

Marcus Terentius Varro (116–27 BCE)

Let us start with Marcus Terentius Varro, a Roman scholar and writer, who appeared to be a significant influencer of his time. He says about Greek writers in his *Res Rusticae* published in 37 BCE: "The Greek writers who have treated incidentally of agriculture are more than fifty in number," and credits every one of them.

Varro cited roses in many of his writings including in *Res Rusticae*— Latin for "agriculture" or also interpreted as "country matters." This work is the often amusing and somewhat humorous dictation of his own experiences and observations of agriculture and farming. They include musings on suburban, commercial flower gardens (see right).

Varro's witty life is mentioned in a study by the Maine Organic Farmers and Gardeners Association: "His wife, Fundania, told him he was no spring chicken (!) Furthermore, he should summarize his knowledge of agriculture (for the rest of us!)." He wrote: "My eightieth year warns me to pack my bags before I set forth on my journey out of life." He lived until he was ninety years old.

Reminiscing about spending time with my own grandmother, Helen, I could imagine sitting with Varro with a cup of tea and listening to him wax poetic about his own horticultural experiences. We can thank Fundania for her encouragement of her husband.

So, in the suburbs of a city it is fitting to cultivate gardens on a large scale, and to grow violets and roses and many other such things which a city consumes, while it would be folly to undertake this on a distant farm with no facilities for reaching the market.

Res Rusticae, VARRO, 37 BCE

Virgil (70–19 BCE)

Influenced by Varro, Virgil's writing of *Georgics*, meaning "agricultural things," references rose gardens of twice-bearing Paestum. His vision of roses was one of maintaining a simple life—not just the highly cultivated garden, but a balance between cultivation and naturalistic views. Virgil tells the gardener:

First he was in spring to gather roses and then apples in the fall, and so he was always first to welcome the swarming bees who make for him the precious honey that was Roman's only sweetener.

Georgics, VIRGIL, 29 BCE

Seneca the Younger (4 BCE–65 CE)

In *An Inquiry into the Natural Sciences* Seneca the Younger relates how roses were cultivated in "forcing houses" and brought on by watering with warm water. Seneca was a teacher and influencer of Nero (see page 90).

Columella (4–70 CE)

Those wanting further advice on roses turned to Columella. Like Pliny, Columella—a Roman landowner and writer of agriculture—treasures the rose for its commemorative roles. Columella writes in the tenth book of his *De Re Rustica*:

Roses, with modest blush suffused, reveal their maiden eyes and offer homage due in the temples of the gods, their odours sweet commingling with Sabaean incense-smoke.

De Re Rustica, COLUMELLA, trans. 1564

In that same work, Columella talks about the "rhythm of the farmer's year" and states that February was the time to plant new rose beds and attend to the old ones; March should see the gardener finished with digging and preparing the late rose bed. He gives planting instructions and offers advice on pruning: "rose tree should be planted

White lilies sown between the furrows in the garden make a brilliant show and the gilliflowers [carnation or similar genus Dianthus] have no less pure a colour; then there are red and yellow roses and purple violets and sky blue larkspur; also the Corycian and Silician saffron-bulbs are planted to give colour and scent to the honey.

COLUMELLA

at the same time as the violet." Like Virgil, Columella includes roses as one of the flowering shrubs favored by bees.

If all his recommendations were followed, the flower garden Columella describes would be a delight to see! He paints the garden, as described above. Columella's words were welcome advice for Romans who were desperate for roses all year round.

Pliny the Younger (61–113 CE)

Another scholar's effort is from Pliny the Elder's nephew, Gaius Plinius Caecilius Secundus, better known as Pliny the Younger. Writing about roses is a familial passion. The Younger, in his *De Re Rustica* (c. 37 CE), details the planting and propagation of roses for the gardener. The Elder (23/24–79 CE), in his encyclopedic *Naturalis Historia* (77 CE), wrote about the advantages of heat in forcing roses to flower early.

Palladius (c. late 4th–5th century CE)

Later, in the fourth century, Rutilius Taurus Aemilianus Palladius (Palladius, for short) wrote fourteen volumes on agriculture and horticulture in his *De Re Rustica*. Palladius advocated the warm-water method for bringing on early blooms. Palladius's writings were also influenced by Varro and Columella.

Emperors' Demand for Roses

Emperors—particularly Nero and Heliogabalus—demanded vast quantities of roses. Thus, large-scale rose production houses were located in Praeneste (modern Palestrina, southeast of Rome), and in Leporia, north of Naples, and Paestrum, farther south, around Egypt's Nile Delta, and at Carthage in Tunisia. Carthage was a center of Roman trade and influence at the time.

In Chapter 3 (see page 65), we made reference to a rose that bloomed a second time in the fall. This "twice-bearing" description probably refers to one rose: the 'Autumn Damask,' that flowered twice a year—the initial bloom and a second sporadic bloom in the fall. This second bloom would be widely prized, as almost all roses only bloomed once at this time. Another explanation for "twice-bearing" is that this possibly could refer to two roses—one that grew early in the season and was forced in greenhouses or grown using warm-water techniques, and another that flowered later—thus increasing the length of the season for blooms. Alternatively, the same type of rose could have been planted at slightly different times of the year, with some plants forced and some held back to prolong the season.

Some of the roses mentioned by Pliny the Elder are the *Rosa gallica*, grown at Praeneste, which flowered late in the season and was most sought after, and a form of *Rosa alba* with white, scented flowers, which was known as the Rose of Campania and grew around Leporia. The "hundred-petaled" rose, previously misnamed as *Rosa centifolia* and now recognized as *Rosa damascena*, also came from the same area.

Despite these significant rose-growing areas in Italy and the various methods described by the scholars to force roses to bloom in fall, demand still outstripped supply. As a result, the Romans imported roses from Egypt and Carthage. It was said that in the fall of 89–90 CE, Paestum produced so many roses that there was no need that year for any to be imported from the Nile Delta. However, the massive Roman cultivation of roses had meant a decrease in their production of corn, as mentioned by Horace (65–8 BCE) in his *Odes*:

O Nile, the Roman Roses are now
much finer than thine?
Your Roses we no longer need,
but send us your corn.

Odes, HORACE

Rosa damascena 'Celsiana' named for a prominent Paris nurseryman, M. F. Cels, who is said to have introduced this rose from Holland. Damask roses are highly valued for their fragrance. This particular variety ages to white, giving a soft, two-color appearance.

ROMAN PLEASURE GARDENS

THE ROMANS EVOLVED their farming practice to include growing plants for ornamentation, and in particular they grew roses for ceremonial purposes. In the ancient region of Latium (which included modern-day Vatican City and Rome), farms produced fruits and vegetables. Cato the Elder (234–149 BCE) suggests that every farm ought to have flower beds and ornamental trees near the house. In his *De Agri Cultura*, written c. 160 BCE, he dictates, "Near the house lay out also a garden with garland flowers."

Later, Columella, in his tenth book of *De Re Rustica*, suggests a garden of a large estate—emphasizing the flowers, which he calls "those earthly stars… the violet beds unfold their winking eye… the rose abashed with modest blush unveils her virgin cheek, the horned poppies with their wholesome fruit and poppies which fast bind eluding sleep."

These farm gardens, packed with fruits, vegetables, and now flowers, offered the farmers opportunities to make more money. In the early days, the ever-popular garlands consisted of chaplets of leaves. Later, crowns of flowers became the norm. Over time, the demand for flowers increased as the

RIGHT *Glaucus and Nydia* (1867) by Lawrence Alma-Tadema is a painting based on Sir Edward Bulwer-Lytton's novel *The Last Days of Pompeii* (1835). Nydia is weaving a rose garland for her dear master Glaucus.

Roman lifestyle became more luxurious. Garlands decorated temples of the gods and, at banquets, roses were placed on tables and used to crown the guests.

Horace, a Roman lyric poet (65–27 BCE), writes: "Let not the roses fail the feast, for lasting parsley, nor the lily soon to die." Horace describes how flower gardens became the indulgence, replacing field crops, thus increased cultivation of flowers continuously enhanced the beauty of the Roman kitchen garden. He further indicates that the rose "had become an emblem of impermanence and decay." However, for all of Horace's distaste of ostentation, garlands played a leading role in the Romans' domestic, religious, and ceremonial lives, marking the rites of passage of birth, marriage, and death. Growing roses for garlands fitted better with Virgil's pastoral mood than with Horace's lack of indulgence, and early writers on agriculture encouraged farmers to grow roses and other flowers expressly for this purpose.

Gardening in Roman Times

Pleasure gardens became fashionable in Italy. Skilled gardeners came into the country from other provinces and allowed the cultured class of Rome to expand the indulgences of gardens in farmlands to converted estates and villas. Examples include the Villa of Diomede at Pompeii and Pliny the Younger's at Laurentum, south of Rome, where Pliny describes the path around his hippodrome (a stadium-like design used for horse racing in Greek times and called a circus in Roman times). In his garden letters, Pliny states: "Farther on there are roses (in the inward sunny circular alleys) too along the path… a visitor on entering would look over the rose-garden, whose scent would reach him."

The details left by Pliny the Younger describe the main features of the pleasure gardens of Rome. The pattern of a formal design that connects the home with the garden is moved forward from the Egyptian and Greek gardens. Moreover, the rose's presence in these gardens confirms its significance in Roman culture. The typical villa had a long portico opening upon a terrace called the xystus (exercise area or path) of the garden. Usually covered with grass, there were also violets, crocuses, lilies, and roses. The outer edges were often lined in boxwood or other shrubs that were trimmed into fantastic shapes. These shrubs were dotted with marble vases or fountains. Examples of this can be found in the wall paintings in the Villa of Livia at Prima Porta and the garden frescoes of Pompeii. The xystus then overlooked a lower garden, called an "ambulation" and designed for strolls, and, finally, a third section, called the "gestation," was enhanced by the provision of shaded avenues for riding on horseback.

Roman millionaires passionately created their gardens, retreats, and villas. Cicero (106–43 BCE), one of Rome's great statesmen and orators, became obsessed with the idea of the city/country life and at one time he was the owner of eighteen estates. The pleasure gardens of the city expanded and surrounded Rome with a circle of green. These estate gardens can be a modern-day study all of their own. Present-day evidence is seen in the Villa d'Este at Tivoli, the Villa Aldobrandini at Frascati, and the Villa Barberini at Rome, all of which occupied the sites of the once-famous gardens of Hadrian, Lucullus, and Nero.

LEFT A red rose shows its stages of bloom: bud, unfurling, and fully open in this fresco from the House of the Golden Bracelet in Pompeii. We can also clearly see its many prickles. Other plants include laurels, poppies, and marigolds.

ROSES: A FAVORITE ROMAN FLOWER

Roses had many uses in Roman life. Horace commented that the fields of Italy were "being transformed into one vast nursery." City officials also worried that roses were a cause for drunkenness. The enormous profits from sales allowed the sellers extra money for drinking—because peasants would haul great panniers of cut blooms into the markets, sell them for big profits, and then head home "well-soaked with wine, with staggering gait and a pocket full of cash."

The stories of the Roman addiction to the rose played out in a multitude of curious ways. Caesar, a famous lover of Cleopatra, is said to have worn rose chaplets to cover his baldness. Another prominent Roman, the governor of Sicily, Gaius Verres (c. 120-43 BCE), traveled around the countryside on a cushion stuffed with rose petals! Cicero (106–43 BCE) was critical of the governor's excessively lavish lifestyle, paid for by heavily taxing the people, and Verres's wearing of rose wreaths and use of rose-petal cushions was critically observed. Stories suggest that roses were also used to cover the smells of the unbathed public, in contrast to these more luxurious uses.

RIGHT In Lawrence Alma-Tadema's painting (1883), Cleopatra meets Mark Antony on her rose-draped barge. She sits in her majestic boat overflowing with flowers, exotic fragrances, and circled by servants and trappings of gold.

CLEOPATRA HAD A LOVE(R) OF ROSES!

Stories of Cleopatra VII (69–30 BCE) carpeting the floor of the banqueting hall or throne room with roses to a depth of about 2 feet (60 cm) when Mark Antony (81–30 BCE) visited, confirm her love of roses. It may have been stories of this occasion that gave rise to the Romans' preference for using masses of roses at their feasts and festivals. Cleopatra is called "The most famous party girl of them all!" in the book *The Untamed Garden: A Revealing Look at Our Love Affair with Plants* by Sonia Day.

Nestling at the heart of the enduring tale of Cleopatra and Mark Antony's passion is the rose. Ensuring her lover would remember her, Cleopatra ordered her slaves to dip the sails of her celebrated barge in rose water to perfume the wafting sea breezes.

Lavish (and Lethal) Uses of the Rose

Who knew that roses could be lethal? The extraordinary lavishness with which roses were used came to light early in the first century. Nero (37–68 CE), a notably ruthless emperor, started the fashion for raining down rose petals from a reversible ceiling on guests at feasts. In the book *Magical Symbols of Love & Romance* by Richard Webster, it is said that Nero spent four million sesterces (about 13 million dollars today) on roses for just one banquet. Nero's obsession persisted—he spent six million sesterces (about 19.5 million dollars) on roses for another party, and, on another occasion, an entire beach at Baiae, near Naples, was strewn with roses at the cost of about two million sesterces (6.5 million dollars). He amused himself by devising different schemes to drop rose petals from the ceiling. It is said that "millions of petals were required for these legendary orgies and guests, having enjoyed a bit of "flagrante delicto" on the couches, actually suffocated under the weight of this fragrant onslaught." What a way to go!

Lucius Apuleius (124–170 CE), who wrote *Metamorphoses*, described the scene of Nero's party: "Venus... heavy with wine and all her body bound about with flashing roses." Book eleven of the ancient Roman novel, *The Golden Ass*, by Apuleius, contains a scene with the Egyptian goddess Isis. She instructs the main character, Lucius, "to eat rose petals from a crown of roses worn by a priest as part of a religious procession." This act helps to regain his human form from that of a donkey.

Many years later, the Emperor Heliogabalus (c. 203–222 CE) wanted to commemorate the start of his reign—at the age of fourteen—with a memorable feast. He certainly fulfilled his ambition. Locking all his guests in the banqueting room to ensure their attention to the spectacle he had devised, he showered them three times with roses. The quantities of petals he used were so overpowering that several people suffocated under their weight. Ironically, roses were also used in abundance at Roman funerals.

All this excess was not without comment. Cato the Elder, a Roman senator and statesman, was an advocate of morals among Romans. He declared that the rose wreaths worn for every minor incident in military life diminished the more accurate sense of their meaning of a victory. Later, Tacitus (56–117 CE), a Roman historian, continued warning of the dilution with roses. He said, "When Emperor Vitellius visited the corpse-strewn battlefield of Bedriacum in northern Italy, laurels and roses were scattered in his path implying that the honor of the

victory was his." The fact was, Vitellius (15–69 CE) did not participate in the fighting. Roses, as symbols of courage and bravery, were being degraded. Thus, the use of roses was like "crying wolf" and began to lose its meaning.

ABOVE *A Banquet in Nero's Palace* is an illustration from *Quo Vadis* (c. 1910). Nero's indulgences included immense amounts of money spent on chaplets and roses for friends. He implemented panels in the ceiling of his dining rooms, that slid back to shower his guests with flowers.

The Roses of Heliogabalus

The painting *The Roses of Heliogabalus* (1888) by Sir Lawrence Alma-Tadema (1836–1912) depicts the ego of the young Emperor Heliogabalus. Roman history describes him as having the worst reputation and leading a disgusting life. He was assassinated at the age of eighteen.

The painter Alma-Tadema was known for depicting Roman life and its excesses. *The Roses of Heliogabalus* depicts the scene as guests are suffocated to death by the enormous weight of millions of roses. Alma-Tadema's inspiration for this painting came from a collection of biographies called *Historia Augusta*, which gives

accounts of releasing violets from sliding ceiling panels. If this story of Heliogabalus is true, perhaps he got his inspiration from Nero.

Roses, Revelry, and Wine

The rose's association with wine and revelry originated during the height of the Roman Empire. Rose chaplets were worn at feasts to cool the brow and counteract the smell of "stale wine." Roses sculpted or painted on a ceiling symbolized secrecy and ensured that whatever was said in the room, while drunk or sober, would be considered confidential.

By the end of the fourth century, Rome's capital had become symbolic at best; other capitals in the country held power. Decimus Magnus Ausonius (310–395 CE), a Roman poet, teacher, and friend of the Emperor, took a period of rest from the politics of the time. In his rose garden, he is credited (as was Virgil) with having written the original version of *Gather Ye Rosebuds While Ye May*. The famous classic poem

I saw rose-beds shining with dew, just like the ones cultivated at Paestum … As long as it is one day, so long is the life of the rose; her brief youth and age go hand in hand… But 'tis well; for though in a few days the rose must die, she springs anew prolonging her own life. So, girl, gather the roses while the bloom and your youth are fresh, and be mindful that so your life-time hastes away.

Translation: On Budding Roses, HUGH G. EVELYN WHITE, 1921

De Rosis Nascentibus evokes the beautiful Paestum rose gardens attributed to Ausonius's comments (shown above). As Ausonius was a sophisticated academic and politician, it is reasonable to expect that his garden contained the same roses of the time as Pliny's. In ancient Roman gardens, these roses are described as *Rosa gallica, Rosa phoenicea, Rosa damascena* (rose of Paestum), *Rosa moschata, Rosa canina,* and the questionable *Rosa centifolia*.

LEFT *The Roses of Heliogabalus* (1888) by Sir Lawrence Alma-Tadema captures the vast quantity of petals with great detail. The young, unscrupulous king watches as his dinner guests suffocate to death under the weight of rose petals released from a retractable ceiling.

PRESERVATION
OF THE ROSE

W<small>E CAN FIND PROOF</small> of the Romans' rampant rose rave in the ruins of Pompeii and its sister city Herculaneum, just outside Naples. When Vesuvius erupted in 79 CE, the daily lives of the Romans who lived there were captured and preserved. Volcanic debris buried Pompeii to a depth of around 13 feet (4 meters) and engulfed Herculaneum to a height of around 21–27 feet (6.4–7.6 meters). The depth of the debris at Herculaneum allowed for the upper floors of the houses to be preserved.

LEFT A section of a garden fresco from the north wall of the House of the Golden Bracelet in Pompeii. Birds, flowers, laurels, and greenery surround a fountain.

In a letter from Pliny the Younger to the historian Tacitus, he recounts the scene:

To a man of such learning as my uncle, this phenomenon (of Vesuvius erupting) seemed extraordinary and well worth investigation. He ordered a light ship prepared, and told me I could come along if I liked. But I said I would rather go on with my work. And, as a matter of fact, he had given me something to write!

From Letter 6.16, translated by
E. M. McCarthy

A few houses have been written about, describing paintings on the walls with floral interest, which include roses. In the case of the House of Venus Marina, it is told the owner cherished the garden so much that after an earthquake damaged his property, he worked on the restoration of the garden before he worked on the house. Frescoes included scenes of doves, fountains, and flowers. The House of the Orchard depicts scenes of many types of trees and the garden there was said to hold roses. In the House of the Golden Bracelet, several frescoes depicting garden scenes are well-preserved, including one with birds and roses. The peristyle court of the common house contained flower beds, vases for growing plants, fountains, water basins, and statues. A great example is the House of the Vettii.

THE DEMISE OF PLINY THE ELDER

It was at the eruption of Vesuvius, on August 25, 79 CE, that we lost Pliny the Elder. He was such an important character in the storytelling of natural history and a great commentator on roses. In *The Story of Garden History*, Penelope Hobhouse states: "The *Naturalis Historia* was an encyclopedic compilation of the known natural world and remained a great dictionary of knowledge throughout the Middle Ages."

A Plethora of Roman Roses

In her book *The Rose: A True History*, Jennifer Potter describes how roses dominated the cities of Pompeii and Herculaneum. Gardens included everything from flowers to vegetables, which were abundant in their frescoes, garlands, and wreaths. The author introduces us to Pompeii archaeologist Wilhemina Jashemski (1910–2007), describing how "Italian flair had turned the Hellenistic peristyle into a 'living, breathing garden'." Archaeologists discovered a commercial flower garden in the Garden of Hercules, which grew scented lilies, violets, and roses, plus olives to provide the base oil in which the flowers were macerated.

At Pompeii and Herculaneum, Pompeian gardeners grew decorative plants for use in medicines and wreaths. The garden of the House of the Chaste Lovers was filled with roses, juniper, and artemisias, among others. More fragrant plants—roses, violets, irises, and olive trees—were found at the House of the Perfume Maker.

Roman Medicinal and Practical Uses of Roses

Just like the Greeks, the Romans employed many kinds of roses as medicine. In 77 CE, Pliny recorded thirty-two different medicinal uses, describing roses as having astringent qualities. Every part of the flower was useful in medication for the head, ears, mouth, gums, tonsils, stomach, rectum, and uterus. The sap of roses was used in ears, as a gargle for gum and tonsil issues and to ease headaches and stomach aches. When mixed with vinegar, sleeplessness, nausea, and fever could also be combated. As Pliny wrote about his rose remedies, he described how crowns of roses would alleviate the "pain in the head" from the wine with "very refreshing effect" (quoted in Potter, *The Rose: A True History*), while parsley, ivy, myrtle, and rose were found to have virtues for dissipating drunkenness.

Rose petals, when charred, were used as cosmetics and to relieve chafed skin. Petals ground to a powder were cooling to the eyes; flowers were known to aid in sleep and would stop discharge with women; and flowers steeped in a drink could help with bowel issues and bleeding. Dried rose seeds were also used for toothache and stomach issues and as a diuretic. Dried rose petals could be used as antiperspirant powders; rose galls when mixed with grease were used for baldness; and *Rosa canina* root was used against rabies hydrophobia. Roses were described as effectively "clearing the brain" when their scent was inhaled.

RIGHT A Roman wall painting of a young woman pouring perfume into a flask from the Villa Farnesina, Rome, c. 10 CE.

ROSES (AND VIOLETS) FOR LEGEND AND ADORNMENT

In ancient Rome, roses were symbols for youth, beauty, desire, fertility, and even an adornment in death. As in Greek mythology, the rose also plays a recurring role in Roman legends. Greek gods and goddesses have their Roman counterparts, and many stories are also told in Roman times.

In ancient Rome, May was the time of the Floralia, an annual festival honoring Flora, goddess of flowers, fertility, and spring. In Latin, the word *flos* means flower, and it is said that the very word "rose" originated with her. Flora, having been struck by one of Cupid's arrows, called out to Eros (Cupid's Greek counterpart.) Her pain caused her to mispronounce the word Eros, instead, sounding "ros." From this, the word "rose" became associated with Eros/Cupid.

Flora is depicted in many works of art, which often show her with roses. In the painting of *Primavera* by Botticelli, Chloris, the goddess of spring and eternal bearer of life, is shown scattering roses on the ground. The painting's pastoral scenery is elaborate, with 500 identified plant species and about 190 different flowers depicted.

Roses Representing Death and the Resurrection

Roses represented both death and the resurrection to come. The festival Rosalia (also known as Rosatio, meaning rose-adornment, or the *Dies Rosationis*, day of rose-adornment) has a specific rose focus. It usually was held in May but was said sometimes to take place later into mid-July. Festivals like this connected the flowers of spring to the rituals of the dead. Violets could also be used in these festivals—their use allowed for the extension of the season of respect. The early bloom of the violets represented the start of the festival season, and the later bloom of the roses signaled the end. The Rosalia festival was further associated with Adonis, Dionysus, and others. The idea also extended into other religions and ceremonies of remembrance.

ROSALIAE SIGNORUM

The military had its own rose festival, Rosaliae Signorum, at which standards were adorned with garlands of roses and violets to remember those lost during battle. The flowers symbolized life and death, while the colors of the violets and roses represented the loss of blood.

LEFT *Primavera* (c. 1482) by Sandro Botticelli depicts 500 identified plant species surrounding the goddess Flora, and 190 different flowers also decorate this scene.

Flowers were placed at burial sites, as it was a custom for Romans to honor their dead. In their religious practice, roses are offered to statues representing a deity or another treasured object. This religious offering is called Rosatio. As the rose is associated with the Egyptian goddess Isis, Romans would honor the goddess in their holiday, Navigium Isidis, which celebrated Isis's influence over the sea, ensuring safety for all sea travelers, and eventually for Roman people. This festival originated in the first century BCE and continued until 416 CE.

Secular flower festivals were condemned when Christianity became the official religion of the Roman Empire. Roses were seen as offensive because of their association with Venus.

If roses were essential to the daily lives of Roman citizens, so too were they in death. The body was adorned with flowers (Rosalia), and a ceremonial meal took place at the tomb. Wealthy Romans left capital sums to pay for continued offerings of food, wine, incense, fruits, and flowers of all kinds, especially violets and roses.

Sprinkle my ashes with pure wine
and fragrant oil of spikenard:
Bring balsam, too, stranger, with
crimson roses.
Tearless my urn enjoys unending spring.
I have not died, but changed my state.

J. M. C. TOYNBEE, 1996

Roman tombs, such as those at Pompeii, were often planted or decorated with gardens and situated on busy routes into the city. Roses could also communicate beyond the grave, as this epitaph by the Roman poet Ausonius suggests (quoted by Toynbee, above).

Though roses were present in funeral rites of the Greeks, in Rome, they were even more associated with death. For a wealthy Roman citizen, preparations of the will would often outline the way roses were to be used for the individual's Rosalia. Funds that had been appropriated for the rose adornment and meetings at funeral sites by Romans throughout the years created a perpetual Rosalia, where the roses represented the continuation of the soul. In the same book we read: "Here lies Optatus, a child ennobled by devotion: I pray that his ashes may be violets and roses..."

The Use of Floral Garlands

Beyond the representation of death and loss, roses and violets were made into garlands and decorated many scenes from drunken festivals to erotic indulgences. Latin references couple violets and roses to a carefree state of pleasure. Before the Common Era, Romans celebrated the end of winter by ritually bathing and offering the first roses of spring to Venus Verticordia, the goddess of changed hearts. Once cleansed, everyone was ready to embark on the new season's round of debauchery.

In both Greece and Rome, the wearing of wreaths and garlands was customary, especially for festive occasions. Garlands of roses and violets were the most popular for Romans and were used for a multitude of occasions, from weddings to drinking parties. The Greek novel *Daphnis and Chloe* (second century CE) mentions the pleasure garden with an abundance of roses and violets.

The Association with Deity

Roses, when associated with gods and goddesses, were a prominent symbol. Among the legends in the Roman era, Aphrodite became Venus, to whom the rose was also sacred. A most famous story surrounds her: as the legend goes, Ouranos was the father of the sky and his wife was Gaia, mother of the earth.

Ouranos and Gaia together created many children, but Ouranos grew fearful that they would one day overthrow him. Ouranos kept all of his children banished under the earth— Gaia's womb. This upset Gaia, who created a sickle for slaying Ouranos and freeing her children. One of their sons, Cronos, took the opportunity to castrate Ouranos and throw his genitalia into the sea, and from that moment Venus/ Aphrodite was born—according to the poet Anacreon, she sprung forth into life from the sea's foam.

This story is celebrated in a fifteenth-century painting, Botticelli's *Birth of Venus*, which depicts light-pink roses growing from where the sea foam fell to the ground. The painting, probably painted in the mid-1480s, portrays the goddess Venus standing on the coast after she had emerged as an adult (in Greek, this translates as *Venus Anadyomene*, the title of Titian's painting depicting the goddess is *Venus is Rising from the Sea*.) Venus is the subject of a multitude of paintings, often including roses, recognized as symbols of love and beauty.

Another Roman story of Bacchus and Cupid tells how Cupid knocked over a bowl of wine with his wing. From this spilled pool of wine grew a rose bush. The rose was consecrated to the Roman goddess Venus as the symbol of beauty.

RIGHT *Birth of Venus* (c. 1485) by Sandro Botticelli depicts Venus born from sea foam and blown by the west wind, Zephyrus, and a nymph, Chloris. The palest pink roses shower the goddess as she arrives onshore, and a handmaid waits to dress her.

There comes a holy and transparent
time when every touch of beauty
opens the heart to tears.
This is the time the Beloved of Heaven
is brought tenderly on earth.
This is the time of the opening
of the rose.

Attributed to Rumi, version by
HAZRAT INAYAT KHAN, 2011

Rosa leucantha
Loiseleur

5

RELIGION AND THE ROSE

(4–1500 CE)

The humanist Greek mentality, in which
groves of trees and roses personified
a deity, was heresy for the early Christian. Any
notion of finding comfort and beauty in a rural setting,
any appreciation of natural scenery as described by
Virgil and Pliny, had vanished. Furthermore, any
legacies that came from the Romans, such as the
irrigation systems and the urban gardens, were almost
forgotten. The concept of gardening for pleasure
is left to the Christian monks and nuns, who had
the leisure and sensibility to grow roses.

EARLY CHRISTIAN ROSE GARDENS

IN STARK CONTRAST to the Rosalia festivals, early Christians did not display flowers at their festivals or tombs. These customs were lost due to their connection to pagan mythology. Roses, especially when woven into crowns and garlands, carried the taint of drunkenness, pursuits of sexual pleasure, idolatry, and unspeakable pagan practices.

The Christian author Tertullian (c. 155–240 CE) and early Christian theologian Clement of Alexandria (c. 150–215 CE) wrote disapprovingly of wearing garlands and crowns. However, around the fourth century, Christians began to disregard these ideas (see right).

Fortunately, what remained for later medieval scholars were Roman gardening manuals, such as those by Varro, Columella, and Pliny the Younger, and descriptions of nature in Virgil's *Georgics*. As a new, more established world emerged, the picture of its gardens can be pieced together

The Christian poet, Prudentius (348–c. 405) 'did not fear to invite his brethren to cover with violets and with verdure, and to surround with perfumes those bones which the voice of the All-Powerful would one day restore to life'

Parsons on the Rose, SAMUEL B. PARSONS (1819–1906)

from written accounts and descriptions in documents, surviving frescoes, and Christian images of monks and nuns—a unique collection of visible evidence. These icons present a rich record of what medieval gardens were like and what grew within them.

While pleasure gardens had almost been forgotten in the 800s, the rise of the rose's importance expanded throughout medieval times to c. 1500. In the 700s, garden roses were grown in southern France. Charlemagne (742–814 CE), Emperor of Rome from 800 CE, and called the "father of Europe," was widely influential. He encouraged rose use, and wrote that roses should be prominently included in Crown land gardens. In fact, he wanted many kinds of plants. In the *Capitulare de villis*, it states:

…it is our wish that they [royal estates] shall have in their gardens all kinds of plants: lily, roses, fenugreek, costmary, sage, rue, southern-wood, cucumbers, pumpkins, gourds, kidney-bean, cumin, rosemary, caraway, chick-pea, squill, gladiolus…

Capitulare de villis, c. 771–800 CE

ABOVE Charlemagne outlines the first rose garden in France. His plant list included numerous herbaceous plants and seventeen perennial trees and shrubs. *Rosa canina* was listed.

The historian John Harvey (1911-1997), a scholar of early horticulture and plant lists, notes the plural use of the word "roses" here, implying that more than one rose variety was grown.

The Benedictines—monks and nuns who follow Saint Benedict (who had his own famous rose garden)—were the first to take the concept of gardens across the Alps. The garden plans of the Saint Gall monastery in Switzerland (a chief Benedictine Abbey, c. 820 CE) are a unique survivor of the time. This amazingly unimpaired artifact shows specific garden areas including sixteen garden beds, featuring a functional area for growing herbs and medicinal plants, as well as lilies and roses. Additionally, eighteen other beds, called the *Hortus*, were for roses (among other plants) grown for altar decoration.

THE RESILIENT ROSE IN RELIGION

PERSIAN GARDENS had four walls, suggesting seclusion and completion. The interior design was of a mandala—a geometric design that holds symbolism for Hindu and Buddhist cultures. The word "paradise" originates from the Avestan word for a garden enclosure or park. The medieval rose garden shares this four-wall design, imitating the idea of paradise.

The gardens that had developed by the twelfth century symbolized the standard image of Eden; rose gardens were built in those that were located in the cloisters of monasteries and convents. Alanus de Insulis (Alain de Lille), a French theologian and poet (1116–1202), described these gardens as an earthly paradise: "a place of the eternal spring, flaming with roses that never wither or die."

As with the Islamic garden for Muslims, the enclosed garden, known by the Latin term *hortus conclusus*, had a sacred meaning for Christians. In both religions, the origin of the idealized garden was the the terrestrial paradise, or Garden of Eden. The symbolic planting and use of the enclosed garden, courtyard, or cloister of the monastery was a metaphor for paradise, as well as for divine and romantic love. Here, roses were widely grown and used medicinally and for their symbolism: red roses for Christ's blood, and white roses for the purity of the Immaculate Conception.

Although the Greeks and Romans focused on the violet and the rose, during medieval times these were replaced by the lily and the rose. The Roman horticultural influences from Varro, Cato, Columella, and Palladius are present here because their manuals and texts survived and were dispersed throughout these monasteries. Gardens of the Middle Ages were an essential part of life. Within these tiny Edens of the monastery, monks could express pleasure in the beauties of nature. The Cistercian Gilbert of Hoyland (c. 1100–c. 1172) writes that a fertile landscape can: "revive a dying spirit, and soften the hardness of a mind untouched by devotion" (Elizabeth MacDougall, *Medieval Gardens*).

Early monasteries developed into a community for therapeutic healing. One heroine is the German Abbess Hildegard of Bingen (1098–1179), also known as Saint Hildegard, who founded the monastery of Rupertsberg (1150) and the monastery of Eibingen (1165). Hildegard is considered to be the founder of natural history in Germany; she wrote: "The earth Sustains humanity. It must not be injured; it must not be destroyed," a relevant statement, especially today, as we face immediate climate concerns. Her writings established categories for plants, dividing them into ornamental, natural, and beneficial. She detailed many medicinal uses of herbs, while her ornamental plants included roses, white lilies, violets, irises, and bay laurel.

RIGHT *The Virgin and the Rose Bush* (1473) by Martin Schongauer shows Mary in an enclosed rose garden, a typical theme of the time. Large, red rose blossoms surround Mary, and a single white rose can be seen close to the Virgin, symbolizing her purity.

ROSE SYMBOLISM IN A RELIGIOUS WORLD

INDIVIDUAL PLANTS became a natural part of the religious calendar, and each monastery would have needed someone to be in charge of ensuring those plants were supplied. Sylvia Landsberg writes in *The Medieval Garden*: "Plants would include bay, holly and ivy at Christmas, yew and catkined hazel to be carried as palm at Easter, birch boughs in May, red roses and sweet-woodruff for chaplets and garlands at Corpus Christi in June, and white lilies and red roses for the feast of martyrs."

The Virgin Mary and the Rose

As Christianity continued to spread, it was easier to adopt previous plant and flower symbols than to eliminate them. From Harold Moldenke's *Medieval Flowers of the Madonna*, we learn that ivy was associated with the earlier myths of Bacchus, and the holly and the

RIGHT An illuminated manuscript has text that is embellished with decorative elements, including initials, borders, and other illustrations. Here, we see the letter C (from Arundel MS) decorated with roses.

yule log were included as part of early Druid rites. All three made their way into Christian festivals in England. The rose, the flower of Hulda (a legendary figure in Germany), is now called *Frau Rose* and "Mother Rose" in England. The rose, which had been given to Venus, and her Scandinavian counterpart, Freya, now became associated with Mary.

By the twelfth century, the Virgin Mary identified as the "beloved" of Solomon. *The Song of Solomon* contains plant allusions, proclaiming, "I am the rose of Sharon, and the lily of the valleys." Mary becomes a garden because of what grows within her; she is an (en)closed garden because she is fruitful for God alone. The Virgin holding a rose could also signify that she is the bride of Christ, as in Fra' Filippo Lippi's *Madonna and Child Enthroned with Two Angels* (c. 1440).

Flowers were associated with Mary some centuries earlier. In the fifth century, Coelius Sedulius, a Christian poet, calls Mary a "rose among thorns" (*Carmen Paschale* II, 28–31). Sedulius's description is perhaps the first interpretation of Mary and the rose. However, the Benedictine monk the Venerable Bede (673–735 CE) had described the lily as Mary's emblem. He said the lily's white petals represented her purity and its golden anthers the glowing light of her soul. Roses, once sacred to Venus, now also became Mary's particular flower. Saint Bernard

(c. 1020–81 CE) described the Virgin Mary as "the violet of humility, the lily of chastity, the rose of charity, the Balm of Gilead, and the golden gillyflower of heaven." The rose *Rosa canina* is specifically mentioned, and red roses generally represented the blood of martyrs.

Mary Gardens

The thornless stem of the rose represents Mary's virginity. During this period, garlands of roses were only permitted to be worn by virgins. Moreover, the Madonna was frequently portrayed as surrounded by roses, as depicted in *The Madonna of the Rose Garden* by Michelino da Besozzo (c. 1425). However, the secular minstrels and poets interpreted the rose symbolically of earthly love—a tradition that continues today.

Specialty gardens started to be developed from the seventh century, which contained plants that were symbolic to Mary. These gardens were called Mary Gardens (or Saint Mary's Gardens), and meant that people could honor Mary in the garden instead of inside the church. According to an article appearing in *The Herbarist*, Saint Fiacre (600–670 CE), the patron saint of gardeners, spent his life caring for a garden at a hospice in France that was dedicated to Mary. Could this have been the garden that later Mary Gardens were modeled on?

Madonna of the Rose Garden (c. 1420–1435) by Michelino da Besozzo or Stefano da Verona. The traditional theme of the *hortus conclusus*—here, an enclosure of roses (symbolic of her virginity)—highlights the Madonna with Child, and also depicts Saint Catherine of Alexandria.

William Johnston's *Encyclopedia of Monasticism* states that the Mary Garden mentioned in the accounting records of Norwich Priory (1373), England, is now believed, after thorough research, to have been a traditional monastic rose garden.

Rosa Mystica

The mystic rose—a story telling of roses woven into three wreaths by the Archangel Gabriel—became descriptive of Mary. Red roses were used to represent Mary's sorrows (as they did for Aphrodite), representing the red of Christ's blood on the thorn bush (instead of Adonis's). White, on the other hand, told of Mary's joy. These stories' similarities to pagan Aphrodite and Adonis of old resulted in Isidore of Pelusium (440 CE) warning: "We should be more careful in marking the difference between the heathen Magne Mater (Aphrodite) and our Magna Mater Mary."

Mary was given many rose names, including Rose of Sharon, the Rose-garland, the Wreath of Roses, Queen of the Most Holy Rose-garden, and the Rose of Modesty (*Rosa pudoris*). The litany of Loreto (Litany of the Blessed Virgin Mary) called her Rosa Mystica (*Rosa mystica*). The "rose without a thorn" symbolizes her purity as the first rose of Eden. According to Christian legend, thorns came to the rose after Adam and Eve were cast from the garden, and Mary became the second Eve. Red and white symbolism is depicted in the biblical epic, *Carmen Paschale* comparing Mary and Eve.

Mary is a rose. Eve was a thorn in her wounding; Mary a rose in the sweetening of the affections of all. Eve was a thorn fastening death upon all; Mary is a rose giving the heritage of Salvation back to all. Mary was a white rose because of her virginity; a red rose because of her charity; white in her body, red in her soul; white in cultivating virtue, red in treading down vice; white in purifying affection, red in mortifying the flesh; white in loving God, red in having compassion on her neighbor.

Carmen Paschale, SEDULIUS, C. 430 CE

ABOVE *The Madonna of the Rose Bower* (c. 1440-1442) by Stefan Lochner shows red and white roses used symbolically.

The Rose in Hymns

Isaiah 11 states: "a staff shall spring forth from the root of Jesse, and a flower shall come up from his root, and the Spirit of God will rest upon him" (Is. 11:11). The rose here is symbolic for Mary. The rose also solidly symbolizes Mary's union between Christ and Mary, in the fifteenth-century English traditional hymn (see right).

There is no rose of such vertue,

As is the rose that bare Jesu,

Alleluia.

For in the rose contained was

Heaven and earth in little space,

Res miranda.

"There Is No Rose of Such Virtue," c. 1420

The well-known Christmas hymn "Lo, How a Rose E'er Blooming," is the English translation of the 1599 German hymn *Es ist ein Ros entsprungen* (also known as "A Spotless Rose"). A line from this translation states that Mary is the "rose that has sprung up to bring forth a child." The connection of the root (vine) of The Tree of Jesse to God is through Mary; it is clear that a flower blooms from the stem. For this, Mary is often referred to as the mystical or hidden rose (mystical can mean hidden), the perfect flower of all humanity.

The theologian John Henry Newman also uses the rose representation in his *Meditations and Devotions* (see below).

Mary is the most beautiful flower ever seen in the spiritual world [...] She is the Queen of spiritual flowers; and, therefore, is called the rose, for the rose is called of all flowers the most beautiful.

JOHN HENRY NEWMAN, 1893

CATHEDRAL OF NOTRE DAME ROSE WINDOW

When the magnificent medieval cathedrals towered above the skyline of ancient cities from the twelfth century onward, many revealed rose windows—artistic circular mandalas decorated with richly colored stained glass. Most were dedicated to Mary.

Perhaps the most famous is the Rose of France window at Notre-Dame de Chartres Cathedral, France, which dates back to 1225. The cathedral is consecrated to Mary, and she is depicted in the famous rose window at the west end of the structure (west was the direction of the Feminine). Mary is seated at the center, carrying the Christ Child, surrounded by angels, doves, prophets, and kings. There is a labyrinth on the floor below the window. If it were possible to fold the wall over onto the floor, the rose window would be perfectly superimposed upon the labyrinth, the center of which is called the rosette.

WRITERS OF ROSES

ALBERTUS MAGNUS (c. 1200–1280), called by many the greatest philosopher and theologian of the Middle Ages, was one of the first to describe a medieval pleasure garden as a place of delight. He was an avid student of Aristotle and his works, and possibly influenced by him and other Roman scholars, he wrote plans of the ideal pleasure garden to include roses (see below).

Within the lawn may be planted every sweet-smelling herb such as rue, and sage and basil, and likewise all sorts of flowers, as the violet, columbine, lily, rose, iris and the like.

De Vegetabilis et Plantis
ALBERTUS MAGNUS, C. 1260

Piero de' Crescenzi (c. 1230/35–1320), adds a section in Book VIII of his 1306 *Liber ruralium commodorum* ("Book of rural benefits") where he talks of "medium-sized gardens" and delves into the topic of pleasure gardens. Crescenzi states, "Once again we see the fenced-in garden enclosed with mixed hedges of fruit trees, thorns, roses, and vines…." Although Crescenzi was said to have plagiarized freely from earlier authors, his work remained one of the essential textbooks for agriculture and horticulture up to and beyond Renaissance times.

In medieval times, the rose garden is also seen as a place for romantic meetings. The rose symbolism within *Roman de la Rose* (c. 1230)—arguably the first modern novel and a tribute

to the Virgin Mary—represents two things: the woman from whom the name "rose" is taken, and female sexuality. Here, again, we see the walled garden.

The Flemish painting *Roman de la Rose* by the master of the Prayer Books shows a noble's pleasure garden of the fifteenth century, complete with a spouting fountain, trelliswork garden divisions, raised beds, turf benches, and roses grown along the railings. These elements of the medieval garden, notably roses trained on a trellis, topiaries, herbs, and seats placed in shady arbors, have endured through the centuries and remain popular in small modern gardens today.

Mary's symbolism has a long history. She is described as "God's rose garden," and Dante announces Mary in his *Paradise* (23:73-74), part of the *Divine Comedy* (c. 1308–1321), with the words: "Rose in which the word of God became flesh." Perhaps the best-known use of rose symbolism can be found in Gustave Doré's illustration of the Celestial Rose based on Dante's work. It depicts Dante and his true love, Beatrice, standing in the heavens staring into the celestial rose, which depicts a multitude of angels and saints. In the thirtieth canto of *Paradise*, the eternal rose is golden. In the thirty-first, it is whiter than snow; the image remains one of the whole of heaven as an infinite, eternal rose, its petals are souls and its fragrance is the never-ending praise of God. In the *Divine Comedy* the mystic rose represents the love of God.

Stories of Roses and Saints

According to legend, Saint Thomas, one of the twelve Apostles, went to the tomb of Mary, not believing the stories of her resurrection. He asked the tomb to be opened and inside found only lilies and roses. Saint Cyprian, who was martyred in 258 CE, and Saint Jerome, who lived in c. 400 CE, praised roses and named them as one of the rewards that martyrs would find in heaven.

Saint Cecilia, patron saint of musicians, was a Roman martyr of the third century who refused to consummate her marriage on the grounds she had committed her virginity to God. Her husband was sent to find proof of this by searching among tombs on the Appian Way. On his return, he found Cecilia talking with an angel, holding crowns of roses and lilies in his hand (see below).

"Guard these crowns with spotless hearts and pure bodies," said the angel, *"because I have brought them to you from God's paradise, and they will never fade or lose their fragrance, nor ever be seen by any except those who love chastity."*

The Golden Legend: Selections,
JACOBUS DE VORAGINE, 1998

Saint Dorothy (fourth century), before her death by execution, was asked by the lawyer Theophilus to send proof of the paradise garden promised by her faith. Moments before her death, she summoned an angel, who delivered a basket containing three apples and three roses, thus proving to the young lawyer the garden's treasures.

In the sixth century, Saint Benedict, who had his own rose garden, *il roseto*, saw in roses only a means of mortifying the flesh whenever he felt the temptation of the world. Maybe this is the early equivalent of "taking a cold shower," as Benedict would cast himself into a thorn bush while naked to escape the temptation of a woman.

Saint Francis is associated with Assisi in Umbria, Italy. One of the largest Christian sanctuaries, the Santa Maria Degli Angeli has an original stone chapel called the Porziuncola, where Saint Francis founded the Franciscan Order. It was here in 1216, that he had a vision in which he witnessed the Virgin Mary. Furthermore, it is also here that the Roselo or "rose garden" exists, which contains a particular prickle-less rose. Similar to Benedict, Francis threw himself into the rose bushes in order to combat doubt and temptation. In contact with the saint's body, prickle-less roses began to bloom and still do today. They have been identified as *Rosa canina* 'Assisiensis.'

Saint Benedict among Thorns (fifteenth century)
by Garelli Tommaso depicts Saint Benedict lying
naked on one side partly covered by a thorn bush,
which he dragged himself through in order to
avoid temptations of the flesh.

Rose Legends

A famous miracle of roses was experienced by Saint Elizabeth of Hungary (1207–1231), the charitable queen. While her husband Louis, King of Thuringia, was away on a crusade, famine swept the country and on returning, he heard embittered tales from his relatives of how Elizabeth had created shortages at the palace by feeding the poor. Louis, believing the tales, banned his wife from her charitable deeds. Nonetheless, she continued to do so secretly, until one day he caught her with a basket full of food. On uncovering it, however,

ROSE EASTER AND OTHER SYMBOLISM

There is an old Pentecostal custom in Rome, going back to the time of Saint Gregory (pope between 590 and 604 CE), who began the scattering of roses on the congregation from the top of the church to symbolize the coming of the Holy Ghost. The custom spread to France and Spain and, as a result, Pentecost is sometimes called "Rose Easter."

As described by the Roman Rites, the rubrics indicate colors of the church for various celebrations during the liturgical year. For Gaudete Sunday (third Sunday) and Laetare Sunday (fourth Sunday) of Lent, the color becomes "rose" (as in wearing rose-colored vestments); the use of the word "rose" (rose pink) rather than "pink" indicates its importance.

By the twelfth century, the rose was incorporated into religious ceremonies, as rose petals were strewn along the aisles. Again, we see the red roses as representing the blood of Christians and white roses the Virgin Mary's purity. According to Christian beliefs, the rose is said to be thornless in the Garden of Eden: Adam and Eve's actions caused the rose to have thorns (prickles) as a reminder to humanity of what was lost. *Rosa sans spina* (*Rosa sans épines*) means without spines/thorns/prickles, representing Mary without original sin.

it proved to contain only perfect white roses, which, since it was wintertime, was rightly construed as a miracle. It is said that even her enemies were converted. A similar story is associated with Saint Elizabeth of Portugal (1271–1336) who carried food for the poor in the German village of Eisenach, just below Wartburg Castle, much against the wishes of her family. While on this journey, she unexpectedly met her husband, who asked to see what she is carrying in her basket. Upon opening her carry, he is surprised to see a bouquet of roses, not food.

Saint Casilda of Toledo's legend is comparable. Saint Casilda was the daughter of the Muslim king of Spain during the Caliphate rule. She would carry loaves of bread for Christian prisoners at the time, hidden in clothes. When caught, the bread would suddenly turn into roses, as portrayed by Francisco de Zurbarán in his painting *Saint Casilda* (c. 1640), and also in *The Miracle of St Casilda* by Zacarías González Velázquez (c. 1820).

RIGHT *Saint Casilda* (c. 1640) by Francisco de Zurbarán depicts the miracle of bread suddenly transforming into roses.

THE ROSARY

THE WORD "ROSARY" developed over time. In the fourth century, Gregory Nazianzus "weaves a chaplet [wreath]" (Anne Winston-Allen, *Stories of the Rose*) for the Virgin Mary. Further along, the daily prayer chain comprised of fifty Aves (roses, according to Latin and German legions) which was used by Catholic Christians to count their prayers developed. The German version describes the "Aves" of roses as a garland with the name "Rosenkranz," later translated into Latin as *rosanum*, "our rosary."

The charm of the rosary can be found with roses. For Christians, the various stages of the rose symbolically represent three images. The rose in bud form represents the infant Jesus; the half-opened rose flower represents Christ's passion; and the fully opened flower represents victory over death. Colors are also symbolic here, representing emotions such as sorrow, joy, and mystery.

Throughout May, Italy and other assemblies use roses to remember the Virgin Mary. Red and white roses are placed on tables, altars, and in the garden. When Saint Dominic instituted the devotion of the rosary, he recognized this symbolism and indicated the separate prayers as tiny roses. May—the Month of May or Madonna's Month—was initially sacred to Flora, Roman goddess of flowers and spring.

THE STORY OF THE ROSARY

A man had the daily desire to create chaplets of roses or other flowers, placing them on the head of an image of Mary. Mary saw the man's enjoyment, devotion, and good intentions, and encouraged him to seek religious life in a cloister. The man's new surroundings and obligations, however, did not allow him any time to make the chaplets, and he was sad. He decided to leave the order and go back to his normal life.

Hearing of the man's dissatisfaction, a priest recommended that he pray fifty Ave Marias in place of making the chaplets. The priest mentioned that Mary would prefer prayers to all of the chaplets he had made throughout his life. Convinced, the man stayed with the order. One day, a request came to the man to run an errand through a forest. While on this errand, the man stopped his horse to recite his fifty Ave Marias. Thieves saw him and decided to rob him and take his horse. As they approached, they saw a vision of a beautiful young maiden standing beside the man, taking 150 roses from his mouth while he was praying and adding them to the chaplet she was making. With her chaplet completed, she placed it on her head and left.

The thieves ran to the man and asked him about the maiden by his side. The man confessed there had not been any maiden and that he was reciting fifty Ave Marias, as instructed. They realized that it was the mother of God who had accepted the rose chaplet in person. The man rejoiced and, from the day forward, returned to making chaplets of fifty Aves for Mary, instructing others to do so. The robbers changed their ways and began a life of good.

RICH ROSE SYMBOLISM AMONG RELIGIONS

DURING the 1200s and 1300s, the European rose story reached a climax. Pagans used roses as decoration to represent their hearts, and Muslims viewed roses as symbols of the human soul. Christians considered roses as a reminder of the Garden of Eden, while Hindus and Buddhists saw roses and other flowers as expressions of spiritual joy.

Rose water and rose oil continue to be used in Islamic rituals in the Middle East and Turkey. One particular pilgrimage to Mecca is known as Hajj. The black cloth of the Kaaba, the "house of Allah," is doused with rose water from Turkey or Iran, and rose oil is used in Kaaba's lamps, releasing the fragrance for all to enjoy. The prophet Muhammad (571–632 CE), the founder of Islam, holds the rose as a symbol. His sweat reportedly smelled of the fragrance of roses.

The story of the "Rose of Tawhid" illustrates the image of a seed planted in the heart of Muhammad. The seed is oneness, striving for expansion; and as it expands, all of consciousness then becomes one.

RIGHT The Rose of Tawhid (also called the Rose of the Prophet) represents the creation from a single point. The Egyptian philosopher Dr. Mostafa Mahmoud (1921–2009) stated: "This concept of Tawheed . . . is actually a verb meaning to make one, that is, to integrate, to bring unity to that which is fragmented."

God created us all from a single soul
(a single point) … A seed planted in
a heart; a heart planted in the soil of
a people, of humanity. A single rose
blossoming. A garden of roses.
One fragrance, one scent, one essence.
The whole journey just for this.

THE QUR'AN

Just as the rose consists
of many petals held together,
so the person who attains
to the unfoldment of the soul begins
to show many different qualities.
The qualities emit fragrance
in the form of a spiritual personality.

HAZRAT INAYAT KHAN (1988)

In Islamic tradition, stories about the rose often characterize it as the flower of heaven. A rose tale about Abraham involves King Nimrod of Urfa (old Edessa) in eastern Turkey. Once, the king threw Abraham into a fire. The fire stopped and miraculously turned into a pond surrounded by roses; thus the fire of earth became roses of heaven. This theme continues as Saint Ali, Muhammad's son-in-law, upon his death bed, requested a bouquet of roses. Once he encountered the fragrance, being satisfied he was to experience heaven, he was able to die.

Washing the Saint Sophia Cathedral in Istanbul with rose water, the Sultan Mehmed II, conqueror of Istanbul in 1453, converted this former Orthodox Christian church into a mosque, known today as Hagia Sophia. It must have taken a huge quantity of rose water to thoroughly cleanse this historic and beautiful place of worship.

Sufism and the Rose

Islam and Sufism embraced the cultivation of geometric gardens—the fragrances within them were seen as representing the sacredness of souls. The Sufi master Jilani is known as "the rose of Baghdad," and his order, Qadiriyya, uses the rose as its symbol. *The Rose Garden* by Sadi and *The Secret Rose Garden* by Mahmud Shabistari by title alone signify the rose's prominence. Hazrat Inayat Khan (1882–1927), a Sufi teacher, relates the soul of a being to the fragrance of the rose. Here, he beautifully describes this teaching (see above right).

The rose and the grail bear similar spiritual qualities. The word "chalice" comes from the Latin *calyx*, which can be defined as "cup." Indeed, the cup-shaped sepals of a flower that support its petals have the same name. Both could be taken to symbolize the soul inviting in divine influence. The rose contains even more symbolism than the grail, stemming from its beautiful design. There is the quantity of the petals, the way they are laid out, and their soft, velvety texture. This is in addition to the rose's intoxicating perfume and its mysterious center hidden within the folds of the petals. A Persian poet from the twelfth century wrote: "mystery glows in the rose bed, the secret is hidden in the rose" (Mara Freeman, *Grail Alchemy: Initiation in the Celtic Mystery Tradition*).

Delightfully, a significant symbol for poets and the Muslim tradition of Sufism is the rose. Sufism emphasizes introspection and spiritual closeness with God. The beauty and purity of roses while placed on thorny branches represents a spiritual path to Allah.

MASNAVI POEMS

A Masnavi is a poem of unlimited length with internal rhyming lines following a meter of eleven (occasionally ten) syllables. Masnavi poems are written in Persian, Arabic, Turkish, and Urdu cultures. They contain a combination of anecdotes and stories taken from the Qur'an, Muslim beliefs, and ordinary tales. These accounts are told to demonstrate a point, and each moral is discussed in detail.

Rumi's *Masnavi* is considered by many as the world's greatest mystical poem. It comprises six books, of which the last was incomplete at his death in 1273. The first two books deal with the deeper self and its sinful inclinations; books three and four have themes of wisdom and understanding; while books five and six deal with the notion that man must reject his human presence in order to recognize God's existence.

symbols with multiple meanings, all of which are implied whenever the words are used. For example, the image of the rose could evoke at once a high spiritual state, a visionary ecstasy, an inspiration, and the rose is that universe. The image of a rose garden embraces all the richness of the created world and all the fantastic splendors of the awakened heart that sees the perfection of God's handiwork emerging in all things and events.

In his works, Rumi writes symbolically about the rose. The agony of love Rumi knew and spoke of so eloquently with the rose is one that is radiant with this "beauty" because it does not merely devastate, it transforms (see below).

Rumi's Rose

The rose is a recurring motif in Rumi's (Mevlana) Masnavi: "Rose is sent to earth by the gardeners of paradise for empowering the mind and the eye of the spirit." The Sufi tradition is that Rumi drew his most profound inspiration from works with a few critical images for centuries—images whose meaning and possibilities of suggestion were seemingly boundless. Chief among these were the "rose" and "rose garden." For the Sufi, "rose" and "rose garden" are

When grace wills, you will know this world to be the Rose Garden of rose gardens, a paradise where nothing is ever born, and nothing ever dies.

Light Upon Light, Inspirations from Rumi,
ANDREW HARVEY, 1996

INSPIRATIONAL QUOTES FROM RUMI

For this author, Rumi has a profound influence. Not only his rose symbolism, but the inspiration of the strung together pearls of words relating to the rose. Here are some wonderful Rumi quotes (among many) involving the rose:

"What was whispered to the rose to break it open last night was whispered to my heart."

"Just this morning, contemplation led me into the rose garden that is neither outside this world nor within it."

"Love is the infinite rose garden."

"Silence you are the diamond in me, the jewel of my real wealth! From your soft earth grow thousands of rose gardens whose perfumes drown me in my heart."

"Love drives you mad from revelation to revelation through ordeal after ordeal until humble and broken you are carried tenderly into the heart of the rose."

"For every house become a window, for every field, a rose garden. Find out of your self, abandon your existence, become me without me."

"My Face is everywhere—and all you are gazing at is roses."

"Stop learning, Start knowing. The rose opens, and opens, and when it falls, falls outward."

The rose looks fair,

but fairer we it deem

For that sweet odor

which doth in it live.

Sonnet 54, lines 3–4,
WILLIAM SHAKESPEARE, 1609

6

THE SECULAR ROSE

(c. 1350–1650)

After about 1350, the rose became prominent in the
secular world. Still-life artists such as Ambrosius
Bosschaert the Elder depicted roses with near-
scientific accuracy, while for writers such as William
Shakespeare and Thomas Campion, the flower was
a frequent symbol in plays, sonnets, and poems.
During the Wars of the Roses (1455–1485) the rival
Yorkists and Lancastrians reputedly wore white and
red rose badges in their fight for the English crown.
Meanwhile, in the Islamic world, roses were important
in the life of the Ottoman court, and Edirne in Turkey
became a great center of rose cultivation.

osa centifolia
liacea

THE ROSE AS A POLITICAL SYMBOL

BETWEEN 1455 AND 1485, white and red roses were the heraldic symbols of two noble houses locked in bitter conflict to win power in England. The struggle—later known as the Wars of the Roses—was between the white rose-wearing supporters of the House of York and the backers of the rival House of Lancaster, associated with a red rose, for the English throne.

Lancaster Red

The red rose in the Lancastrian heraldic badge derived from one worn by King Edward I of England (r. 1272–1307) in the fourteenth century. Edward had a golden rose badge and his brother Edmund Crouchback, the first Earl of Lancaster (1245–1296), adopted the red rose as his emblem. The rose in question is believed to have been *Rosa gallica*, also known as 'Old Red Damask' or the 'Rose of Provins,' which grew wild in central Asia. It was known to

Persians and Egyptians in antiquity, and was cultivated by the Romans, who introduced it to their province of Gaul (hence the name *gallica*).

The white rose in the Yorkist badge, meanwhile, derived from a badge used by Edmund of Langley, 1st Duke of York (1341–1402). As we have seen in the previous chapter, the white rose connected symbolically to the Virgin Mary, mother of Jesus, often described as the "Mystical Rose of Heaven."

Rose of Provins

The House of Lancaster

The royal House of Lancaster was established by Henry Bolingbroke, who deposed King Richard II and seized the throne as King Henry IV in 1399. Henry was the son of John of Gaunt and Blanche of Lancaster, heiress to the Duchy of Lancaster as daughter of Henry of Grosmont, 1st Duke of Lancaster (c. 1310–1361). John of Gaunt was the fourth son of King Edward III and Henry claimed the throne through this royal descent— as grandson of Edward III.

The House of Lancaster ruled for the reigns of three kings: Henry IV himself (r. 1399–1413); his son Henry V (r. 1413–1422), the celebrated victor in the battle of Agincourt in 1415; and Henry VI (r. 1422–1461; 1470–1471), who came to the throne at the age of just nine months (the youngest ever king of England) but proved an ineffective ruler whose decline into madness saw the rise of Richard, 3rd Duke of York. Richard became Protector and Defender of the Kingdom in 1454.

The Wars of the Roses

Richard had a feasible claim to the throne himself. He was the great-grandson of King Edward III via Edmund of Langley, 1st Duke of York. The struggle between supporters of the rival Lancastrian and Yorkist claims to the throne is what constituted the Wars of the Roses.

Some historians claim that while the white rose badge was definitely used by Yorkists in these wars, the idea that the Lancastrians had red rose badges is a later invention. It's certainly true that the conflict was not known as the Wars of the Roses until the nineteenth century, when Scottish novelist Sir Walter Scott described the struggle in those terms in his novel *Anne of Geierstein* or *The Maiden of the Mist* (1829).

This illustration by A. S. Forrest reveals how rival nobles picked white or red roses to signal their allegiance to York or Lancaster—as Shakespeare described in *Henry VI, Part 1*.

Nevertheless, the association of the red rose with the Lancastrian cause certainly had contemporary meaning and currency, for the connection was made by William Shakespeare in his play *Henry VI, Part 1*, written c. 1589–1592. Moreover, when Henry Tudor became King Henry VII in 1485 and married the Yorkist heiress Elizabeth of York, bringing the Wars of the Roses to an end, he created the Tudor rose, combining the white rose and the red to symbolize an end to the years of conflict and a unified future.

The White Rose of Elizabeth I

Henry VII's granddaughter Elizabeth I reigned for 45 years (r. 1558–1603) but never married. This was a time of vast and wonderful achievement for England, when the mighty Spanish Armada was repelled (1588); explorers such as Sir Walter Raleigh (1552–1618) traveled to the New World, and claimed parts of it for the crown; seamen like Sir Francis Drake (c. 1540–1596) circumnavigated the entire world and demonstrated England's might on the seas; and at home poet-playwright William Shakespeare (1564–1616), poet Edmund Spenser (1552–1599), composer William Byrd (1543–1623), artist Nicholas Hilliard (c. 1547–1619), and others were part of a golden age of artistic expression in England that arguably has never been matched.

The queen was celebrated and, because she never married, a cult was fostered around her as the Virgin Queen, a ruler who did not need to take a princely husband because she was "married" to her people. She declared near the start of her reign in 1559, "Nothing, no worldly thing under the sun, is so dear to me as the love and goodwill of my subjects" and that "in the end it shall be for me sufficient that a marble stone shall declare that a queen, having reigned such a time lived and died a virgin."

At the same time, the language of Arthurian chivalry and courtly love was used at Elizabeth's court, according to which the queen was an all-powerful mistress who granted or withheld favors to petitioning admirers, so lording it over some of the most powerful men in the country. Aware of the white rose's religious association with the Virgin Mary, and its secular association with love and early beauty, she adopted it as one of her emblems. Portraits often show Elizabeth with a white rose—and with pearls, another traditional symbol of virginity.

THE ROSE IN ART AND ARCHITECTURE

THE TUDOR ROSE created by Henry VII consisted of a white rose set within a red rose. It was widely used as a symbol of unity in architectural detail, paintings, and embroidery. It came to be called "the flower of England" and today is England's national flower. There are splendid examples of the Tudor rose in King's College Chapel in Cambridge and at Hampton Court Palace. It appeared often in Tudor portraits, including the celebrated Pelican Portrait of Elizabeth I painted by the great Nicholas Hilliard in 1575.

The Tudor rose

Across the Channel, roses often appeared in Dutch still-life paintings of flowers—and these and other flower paintings are among the best sources of knowledge of the roses cultivated in early gardens.

RIGHT The Tudor rose—white within red—appears top left in this 1575 portrait of Elizabeth I by Nicholas Hilliard. It symbolizes the royal dynasty established by Elizabeth's grandfather, Henry Tudor, and the rose's medieval association with the Virgin Mary is suitable for a ruler known as "the Virgin Queen."

The Rose in Still-life Paintings

Ambrosius Bosschaert the Elder (1573–1621) was one of the first artists to specialize in painting still-life images of flowers. Born in Antwerp, he fled to the northern Netherlands with his family as Protestant refugees in c. 1587. Painting on copper and rendering the flowers and fruit with a scientific level of accuracy, he made precise still-life images of fruit and flowers in carefully grouped compositions that balanced color and form.

His *Still Life of Flowers* (1614), for example, delineates a yellow tulip, white rose, and pink carnation in front of brightly colored flowers in a basket.

In this and many of his works he depicted a variety of flowers that do not bloom at the same time—so it would be impossible to create this floral display in real life. He was immensely successful: in 1621, he was paid the vast sum of 1,000 guilders (the equivalent of around 42,000 dollars today) for a flower painting commissioned by the Prince of Orange's chamberlain. He died in that same year, but his sons Ambrosius Bosschaert II (1609–1645), Johannes Bosschaert (1606/08–1628/29), and Abraham Bosschaert (1612–1643), and his brother-in-law Balthasar van der Ast (1593/94–1657) were also artists. Van der Ast's *Basket of Flowers* (1622), *Chinese Vase with Flowers, Shells and Insects*, and *Flowers in a Vase with Shells and Insects* (both c. 1628) are compositions featuring beautifully rendered roses and other flowers.

Roses in Religious Art

In the Christian as in the Islamic world, the rose was used in medicine. And because of its powerful religious associations it was cultivated in monastery gardens (see page 115). During the fourteenth to seventeenth centuries, roses also continued to play a significant role in Christian religious art and iconography.

The *Ghent Altarpiece*, a masterpiece painted by Hubert (c. 1385–1426) and Jan van Eyck (c. 1390–1441) and completed c. 1430, features in its upper central panel the Virgin Mary wearing a floral crown containing lilies, roses, lilies of the valley, and columbines— with roses standing for love, lilies of the valley for happiness, lilies for virginity, and columbines for humility. Rose bushes flank the *Adoration of the Lamb* in the central lower panel.

In another van Eyck masterpiece, the magnificent *Madonna at the Fountain* of 1440, rose bushes again frame the subject, the Virgin Mary holding the Christ child before a banner held aloft by angels. Here the rose bush, along with the beads held by the Christ child, suggest the rosary used by Catholic Christians to count their prayers.

Another exquisite representation of the Madonna with roses is the *Madonna of the Rose Bower* (c. 1440– 1442; see page 122) by German artist Stefan Lochner, in which an abundance of red and white roses behind the seated Virgin symbolize her purity and innocence.

Rose Miracles

Many miracle stories told of holy women featured a miraculous appearance of roses (see pages 126–9). One of these was the rose miracle that reputedly occurred in December 1531 in Mexico, when on December 9 the Virgin Mary appeared to a native laborer named Juan Diego in what is now Mexico City and instructed him to go to the local bishop Fray Juan de Zumárraga with the message that Mary hears the weeping and sees the sorrow of the poor and will improve their conditions—and asking him to build a chapel in her name. The bishop refused to believe what the laborer said and sent him away. That same day Juan saw the Virgin a second time. She instructed him to keep trying.

When Juan visited the bishop again the next day, December 10, the cleric again refused to accept the message and told Juan to ask the Virgin for a miraculous sign to prove her identity. The next day, December 11, Juan did not visit the hillside because his uncle was seriously ill and early on December 12, with his uncle seemingly about to die, Juan set out to find a priest to hear his uncle's confession and give him the Last Rites. He tried to avoid the hillside because he was embarrassed at having failed to go and meet the Virgin the day before; but she intercepted him on his journey and when he explained what was wrong, she gently told him off for not having come to her for help. She said to him, in celebrated words that are now carved above the door of the Basilica of Guadalupe: "Am I not here, I who am your mother?" She told him that his uncle had recovered. Then she showed him beautiful Castilian roses that were growing nearby—despite the fact that it was midwinter and freezing cold. After he picked the flowers, she laid them out in an intricate arrangement within Juan's cloak so he could walk along carrying them. Finally, the archbishop accepted the message—and when Juan opened his cloak the flowers fell to reveal a vision of the Virgin.

RIGHT *The Miracle of Roses* (c. 1735–1740) by André Gonçalves is another depiction of the bread to roses miracle. Here, Queen Elizabeth of Portugal had bread hidden but when her husband demanded to see it, roses appeared (see page 129).

The rose was also symbolic of youth and youthful beauty. The narrative poem *Venus and Adonis* (published 1593) describes "rose-cheek'd Adonis' going hunting" (line 3). In the history play *Antony and Cleopatra*, probably written 1603–1607, the rose symbolizes youth and its passing. As a flower blooms beautifully, then fades and decays, so people flourish in youthful beauty and perhaps become powerful as a result, then fade in older age and lose their position. Antony refers to "the rose of youth" (3.13.20), while Cleopatra bemoans her lost beauty and power with the image of a blown rose (one that has gone over and is dying). She says: "See, my women, / Against the blown rose may stop their nose / That kneeled unto the buds." (3.13.38–40). She means to say: her servants hold their noses to avoid the smell of a dying rose but they were happy to inhale its fragrance when it was a bud.

The thorns of the rose bush featured in *Venus and Adonis*, where the poet likens the persistence of a lover to a person picking a rose, despite the fact that thorns might put them off (see right).

The rose also appears as a comparison in the tragedy *Romeo and Juliet* in a line that was so often quoted it has become proverbial: "a rose by any other name." The star-crossed lovers, Romeo and Juliet, bemoaning the fact that because of their names, they cannot be together—their families, the Montagues and Capulets, are locked in a bitter feud in Verona, Italy. Juliet declares: "'Tis but thy name that is my enemy" (meaning we are only enemies because of our names), and adds: "What's in a name? That which we call a rose / By any other name would smell as sweet" (2.2.43–44). Her meaning is that a rose does not smell sweet because we call it a rose: it could be called a tulip or a nettle and would still smell sweet. To say something was "sweet as a rose" was a proverb popular in Shakespeare's time.

*Foul words and frowns must not
    repel a lover;
What though the rose have prickles,
    yet 'tis pluck'd:
Were beauty under twenty locks kept fast,
Yet love breaks through and picks them
    all at last*

Venus and Adonis, 1593 (lines 573–576),
WILLIAM SHAKESPEARE

Other Poets' Symbolic Use of the Rose

Poet and composer Thomas Campion (1567–1620) wrote to his beloved in his poem *My Life's Delight*: "Thou all sweetness dost enclose! / Like a little word of bliss:/ Beauty guards thy looks. The rose / In them pure and eternal is."

For poet and dramatist Christopher Marlowe (1564–1593), roses made a perfect bed for intimacy in his celebrated poem *Come live with me and be my love*: "Come live with me and be my love … / … And I will make thee beds of roses / And a thousand fragrant posies."

For the seventeenth-century poet Richard Lovelace (1617–1657), also, roses were the flowers of love. He wrote from prison to his beloved in 1642 (see right).

When love with unconfined wings
Hovers within my gates,
And my divine Althea brings …
… When I lie tangled in her hair…
When flowing cups run swiftly
*   round,*
With no allaying Thames,
Our careless heads with roses bound,
Our hearts with loyal flames.

To Althea, from Prison, RICHARD LOVELACE

In the poem *The Rose*, Lovelace calls on the flower to fly to the room of his beloved and fill it—to cover the floor and the couches and bestrew her hair (see below).

See! Rosie is her Bower,
Her floore is all this Flower;
Her Bed a Rosie nest
By a Bed of Roses prest.

The Rose (1642–1646), RICHARD LOVELACE

He addresses the rose as "sweet serene skye-like Flower," "Vermilion Ball that's given / From lip to lip in heaven" and as "Dear Offspring of please'd Venus and sweet, plumpe Silenus," referring to the Roman goddess of love, Venus, and Silenus, a companion and tutor of the wine god Bacchus.

Gerard's Herbal

English botanist John Gerard (c. 1545–1612) made some interesting comments on the Great Holland Rose in his book *Herball, or Generall Historie of Plantes*. Gerard, who kept a large herbal garden in London, mainly based the 1,484-page illustrated guide on an English translation of a previous herbal published in 1554 by Flemish physician and botanist Rembert Dodoens (1517–1585). Gerard's edition, first published in 1597, became very popular and was widely used in England in the seventeenth century.

He called the rose "*R. damascena flore multiplici*, the 'Great Holland Rose,' commonly called the 'Province rose.'" This rose—now known as a *Centifolia* 'Gros Choux d'Hollande'—was certainly connected with the Netherlands for it often features in seventeenth-century Dutch floral paintings. He not only described the plant but also gave a detailed description of its medicinal uses.

Gerard wrote that rose water was good for "strengthening of the heart, & refreshing of the spirits." He also described its use for eye inflammation problems: "paine of the eies proceeding of a hot cause," and for treating constipation: "The juice of these roses… doth move to the stools, and maketh the belly soluble." "Juice" made from musk roses and Damask roses was particularly effective, he noted. He added that rose syrup (called in Latin *Serapium*) was a calmative: it could be used to treat fevers and settle internal inflammation ("the inflammations of the intrails"). He observed that the same effects could be achieved with sugared rose petals heated over the fire. Rose water was a cure for insomnia—it "bringeth sleep," he wrote—and was good "for all things that require a gentle cooling."

USE OF ROSE PRODUCTS IN ISLAMIC MEDICINE

Meanwhile, in the Islamic world, roses and rose products were often used in treating ailments. In the fourteenth century, Ishak bin Murat's book *Edviye-i Müfrede* ("Simple Drugs") described the use of rose powder—made by crushing petals with a pestle and mortar—to treat the skin. The preparation reputedly cleared pimples and was used to treat scabies. It was found in Turkish baths.

In the thirteenth century the Andalusian physician, botanist, and pharmacist Ibn al-Baitar (1197–1248) wrote of the calming effect of rose water and recommended its use for slowing rapid heartbeats caused by anxiety. He suggested boiling rose water and allowing the steam to envelop the head to treat headaches: he noted it was good for treating eye diseases. He wrote: "Rose water strengthens the mind and the brain, sharpens the senses, increases the life force."

By this point, Islamic physicians and botanists had been recommending roses, rose water, and other rose products for centuries. As early as the ninth century CE, physician-philosopher Al-Kindi (801–873) recommended rose products for treating diseases of the liver and mouth as well as stomach pain and ulcers. Around the same time, the polymath Abu Hanifa Dinawari (815–896), known as a botanist, astronomer and historian among other things, recommended using rose water for treating fever.

In the eleventh century Ibn Sînâ (980–1037 CE), the physician and astronomer known in the West as Avicenna, noted the beneficial effects of roses and rose fragrance on the brain and heart. He said the fragrance was calming and "highly beneficial for fainting and for rapid heart beats." He also believed that rose fragrance helped brain function, boosted concentration, and improved memory. Avicenna's ideas were influential for centuries. His five-volume medical book *The Canon of Medicine*, completed in 1025, became a standard medical textbook in the West after it was translated into Latin in the thirteenth century and was used right through to the eighteenth century.

For Avicenna and some other writers, rose water and rose fragrance had a spiritual as well as a physical benefit. Avicenna wrote that because of its wonderful fragrance: "the rose addresses the soul." In his fifteenth-century medical book *Kemaliye*, Mahmud of Shirvan described the practice of applying rose powder to the body after bathing; he said this not only made the body fragrant but also eased the spirit. He said the fragrance of roses was like the fragrance of angels.

THE ROSE AT THE OTTOMAN COURT

ROSE WATER AND ROSE PRODUCTS including rose syrups, rose sherbets, and other desserts were widely used at the court of the Ottoman sultans. Courtiers served rose water at banquets and during meetings of high-ranking officials. Rose water was also a key ingredient of the celebrated miski (perfumed soaps) used at the Topkapi Palace in Istanbul and in the desserts made in the Helvahane (halva kitchen) there. Rose water and bottles for rose water were the most popular items on official lists of gifts offered to the Ottoman sultans.

Large gardens close to the Topkapi Palace near the Marmara Sea included an area called the Gülhane Gardens for growing roses. The word *gülhane* means "House of Roses," from the associated Persian *Gulkhana* ("house of flowers"). Palace attendants gathered roses each spring to make rose water and other rose products on site. The products were also purchased from other sources because vast amounts were needed:

in 1642, according to official records, palace officials bought 2¼ tons (about 2,000 kg) of rose water.

The facilities in Istanbul could not produce enough rose water for the needs of the imperial household. Substantial quantities of rose saplings and rose water were imported from Edirne (formerly known as Adrianople in northwestern Turkey)—the main center for growing roses and making

ROSE WATER FIRST

Reputedly after Sultan Mehmed II "the Conqueror" (r. 1444–1446; 1451–1481) transformed the Hagia Sophia Church in Istanbul into a mosque he required that it be washed with rose water before it could be opened (see page 134).

rose water in the Ottoman territories. The rose saplings sent to Istanbul were grown in the palace gardens there.

At Edirne, there was a *gülhane* associated with the royal palace. The polymath Kâtip Çelebi (1609–1657), author of geographical encyclopedias among other works, described how the city had 450 rose gardens along the banks of its three rivers. He said that the rivers overflowed and flooded the adjacent lands at the end of winter.

Traveler Evliyâ Çelebi (1611–1682) wrote: "there is no country in all the lands of Anatolia with such fertile soil, extending to all corners" and listed the great abundance of flowers growing there, most of all hyacinth, sweet basil—and roses.

RIGHT *Mehmed II Smelling a Rose* (a miniature from the Topkapi Sarayi Albums c. 1481). The rose here represents the Sultan's love of culture.

Making Rose Water

Roses and rose water were not only used by the elite in the imperial household. According to Evliyâ Çelebi, there were fourteen rosewater shops in the Old Bazaar in Istanbul in the 1640s, employing seventy people. He described how female sellers from Edirne sold rose water from large copper cauldrons situated in front of the bazaar.

There was an established tradition of growing roses and making rose water in Anatolia. As early as the thirteenth century, Andalusian botanist and

BELOW *The Perfume Maker* by Ernst Rudolph shows workers carrying roses in baskets and removing the petals by hand before dropping them into large ceramic pots for distillation.

pharmacist Ibn al-Baitar noted that the roses from Nisibis (now Nusaybin in Mardin province, Turkey) had the sharpest fragrance, and they were made into rose water.

Moroccan traveler and scholar Ibn Battuta (1304–1369) reported that roses were grown and rose water made at Nisibis. He wrote that rose water made in that area had "a unique fragrance and a unique taste." Arab geographer Al-Dimashqi (1256–1327) also cited Nisibis—alongside places in what is now Iran, notably Firuzabad and Quwar—as major centers of rose water production, where the water was bottled and sent out by ship around the Muslim world.

ROSE WATER IN THE MIDDLE EAST

Many tales were told about how people first discovered and developed uses for rose water. In one, making and using rose water developed from the practice among Turkish tribespeople of washing sacrificial horses and other animals with scented water. Roses grew wild and were cultivated in the great city of Constantinople, and rose water and other rose products were important in the customs of the Byzantine emperors there, just as they were later at the Ottoman court.

Large quantities of rose water were also use medically. According to an official document of 1489, the Edirne Darussifasi (hospital), founded in 1488, had three lead furnaces for making rose water.

Rose water also has a significance in the Persian New Year celebrations, during which family members and visitors are blessed with a sprinkle of rose water from a silver container called a *gulabaksh*. The water is sprinkled and rubbed on the palms of the hands and the face, which has a spiritual significance, as the rose heals painful emotions from the heart chakra, replacing them with peace, harmony, clarity, love, and joy.

*From what invisible rose
garden was flung this rose
whose perfume maddens me
and makes me lucid?*

RUMI, 1010, IN, ANDREW HARVEY, 2007

Rosa moschata plena

7

STOP AND SMELL THE ROSES

(1650–1789)

For many centuries, we have learned from ancient philosophers, writers, and botanists about the pleasures, medical benefits, and rituals of rose water and rose oil. Let us add to these stories the subject of the undeniable, treasured fragrance of the rose. A recent study suggests that there is a scientific reason to "stop and smell the roses" (a phrase credited to Walter Hagen in *The Walter Hagen Story*). It claims that people are happier when they take the time to appreciate the good things in life, as Adler and Fagley observed: "acknowledging the value and meaning of something—an event, a behavior, an object, and feeling a positive emotional connection to it."

THE FRAGRANCE OF THE ROSE

Rosa bifera officinalis, known as the perfumer's rose.

OVER AND OVER, rose lovers claim memory to the scent of the rose—as, for example, a nod to an earlier time or (for me) a reminder of my grandmother's home. As Jonathan Reinarz says in *Past Scents*:

Smell is a fleeting sense, the moments we sense a smell it turns into a past scent. But at the same time, when you smell these things again later in life, they immediately conjure up very vivid memories.

Reinarz, 2014

The Subjectivity of Fragrance

Fragrance can be personal. A person and a companion may stroll into the rose garden, be inspired by a particular bloom, bend over to allow their olfactory senses to be wooed and swayed, and say: "Oh! This rose smells incredible! It reminds me of when I was growing up." However, their companion might smell that very same rose and declare: "I don't smell anything!"

THE SCENT OF THE 'AUTUMN DAMASK' ROSE

Robert Calkin (1999) describes the *Rosa damascena* 'Quatre Saisons Continue' ('Autumn Damask,' before 1633) as one of "the most beautifully scented of roses," saying, "if sunshine had a smell, this would be it!"

He describes the spiciness of the fragrance from the seed parentage of *Rosa moschata* (musk, spice, clove, and fruity, akin to banana); and the pleasant old rose quality fragrance from *Rosa gallica* (pink pepper, incense, rose and myrrh, wood notes). The pollen parent is *Rosa fedtschenkoana*, a species rose discovered c. 1871 by Alexei and Olga Fedtschenko. *Rosa fedtschenkoana* is native to central Asia and northwestern China and has a surprisingly strong linseed (flaxseed) oil-like fragrance. Flaxseed is a nutritional supplement, and is sometimes used in cooking. Its scent has a nutty aroma—a perfect example of the vast range of rose fragrances.

In a 2015 interview, Michael Marriott of David Austin Roses Ltd. compared the fragrance from *Rosa fedtschenkoana* to: "a little bit of blackberry jam on Hovis bread!" An example of the subjectivity of fragrance at its best!

Rose authors say that the original 'Quatre Saisons' is probably extinct. However, the 'Autumn Damask' was reintroduced into cultivation in 1959 by Graham S. Thomas, who found it as a sport of the 'Perpetual White Moss.'

Rosa damascena 'Quatre Saisons Continue'

The Sophistication of Scent

Fragrance is complex; a point made clear by doctors Linda B. Buck and Richard Axel, who were deservingly awarded the Nobel Prize for Medicine for their work in 2004. Their research states that we have approximately 1,000 genes that are coded for types of olfactory receptors. Intricacies of fragrance are also brought forth by Victoria Henshaw, author of *Urban Smellscapes*. Even though we now know 1,000 genes are working for our sense of smell, Henshaw explains that we are born with very few smell preferences, which we learn as we grow older; perhaps based on our culture, where we grew up, and our exposures.

Carl Linnaeus (see page 12) is not only the classifier of plants; in the eighteenth century he organized odors into seven broad categories: "aromatic, fragrant, musky, garlic-y, goat-y, fetid, and nauseous." Neuroscientist Jason Castro (Bates College, Maine),

developed similar categories: "fragrant, woody, fruity, lemony, minty, sweet, popcorn, chemical, pungent, and decayed." Guiding us to the odors of pleasant smells alone, the English perfume expert, Eugène Rimmel (1820–1887), known as Britain's first historian of perfume, counted as many as eighteen pleasing categories.

George William Septimus Piesse (1820–1882), a chemist and optician, attempted to categorize odors in *The Art of Perfumery: And Method of Obtaining the Odors of Plants*. Piesse created what is called the odophone: a system consisting of perfume atomizers that were activated by the keyboard, or keys of the piano. The lowest "note" of the register was the fragrance of patchouli, the top note was the fragrance of the rose, and in "playing these notes, Piesse 'composed' perfume" (Amanda Smith, "The Modern Science of Smell"). Let us call that musicology with a rosy harmony!

RIGHT Sometimes known as *My Sweet Rose*, *The Soul of the Rose* (1908) is by John William Waterhouse; the artist took inspiration for his works from tales of romance. Here, we see the subject smelling the rose in a way that the viewer can easily understand.

The "Moss" class of roses has bloom characteristics and
fragrances similar to those of previous rose classes. But
the addition of "mossy" glands on the buds distinguish
this class of roses from its Centifolia parents.

Variations in Fragrance

In Chapter 1, we discovered the wide variety of rose classes, their colors, and flower forms. It may be a surprise that there are the same numerous variations in fragrances as there are in roses.

Roses can reveal (to name just a few): old rose, tea, myrrh, and musk scents. We can also be surprised to find fruit aromas such as apple, banana, blackcurrant, all citrus, guava, lychee, pear, raspberry, and strawberry.

TWO FRAGRANCES FOR THE PRICE OF ONE

Rosa rubiginosa (also known as *Rosa eglanteria*) is a species of rose native to Europe or western Asia; it is called "Sweet-Briar," because it releases a sweet (sometimes fruity) smell on a warm day, released by a gentle wind or light rain. The rose bloom itself has a natural rose floral scent, but if you press the lower parts of the leaves you will experience the fragrance of green apples! One of the great references for this rose comes from Shakespeare's *A Midsummer Night's Dream* (1595–1596), where Oberon describes Titania's bower:

I know a bank where the wild thyme blows,
Where oxlips and the nodding violet grows; as it is set out like this
Quite over-canopied with luscious woodbine,
With sweet musk-roses and with eglantine.

(2.1. 235–239)

The Moss class of roses is the genetic mutation (sport) of the Centifolia class. The flower buds are covered with glandular tips that produce their own oily (sometimes sticky) scent, which can range anywhere from citrus, pine, or anise, to earthy notes, making an excellent complement and contrast to the rose scent of the flowers. Touch the moss and get one scent on your fingers; then, smell the rose to get the floral essence—two for one.

'Kazanlik' and the Composition of Fragrance

Relying on the expertise of those with a knowledge of fragrance, it is interesting (yet not surprising) to learn that the chemical composition that makes up the rose's particular fragrance is as complex as the rose variations themselves. Robert Calkin ("The Fragrance of Old Roses") uses *Rosa damascena* 'Kazanlik' (before 1612) as an example. Famous for its delicious fragrance, 'Kazanlik' is widely used in the production of rose oil, rose water, and the perfume industry.

Calkin explains that 'Kazanlik's' scent reveals some 400 known separate ingredients. Just think about that for a minute. In his book, *Perfumery, Practice and Principles*, Calkin continues to describe that varieties of roses all have different scents from one another. Here are a few examples:
• 'Lady Hillingdon' Tea (1910): strong tea with apricot
• 'Madame Isaac Pereire' Bourbon (1876): intense raspberry, rich like ripened fruit
• 'Belle de Crécy' Gallica (1829): strong sweet rose
• 'Madame Alfred Carrière' Noisette (1875): sweet tea
• 'Ispahan' Damask (before 1827): rose with gourmand (honey, chocolate, vanilla), balsamic, woody, patchouli, and musky notes.

In the class of *Rosa gallica* and the classes of roses that followed, Damasks, Centifolias, Albas, etc. all have significant fragrance components called rose alcohols. These alcohols are identified as phenylethyl, citronellol, geraniol, and nerol. Each of these alcohols occurs in different percentages in every rose. Describing these alcohols further (quoted material from Calkin, "The Fragrance of Old Roses"):
• Phenylethyl alcohol has a "soft rose petal-like" character and is the main ingredient of rose water.
• Citronellol has a "wonderfully warm and vibrant" role, grassy and citrus-like (think citronella candles) and is displayed by some of the Rugosa hybrids.
• Geraniol alcohol has a sharper floral "geranium leaf" rose quality. It is said to lift, enhance, bind, and modify odors.
• Nerol is described as the "harshest of all these previous, and fresher"—a sweet rose with a bitter citrus twist.

All of these rose oils together make up the old rose fragrance that is so detectable and familiar to our noses; each of these oils on their own, however, maybe not so much.

RIGHT It is the combination and complexity of the 400 ingredients in the rose 'Kazanlik' that together delivers its magnificent scent.

THE BULGARIAN ROSE OIL INDUSTRY

THE NAME 'KAZANLIK' comes from the city of Kazanlak, at the foot of the Balkan mountain range, where the Bulgarian rose oil industry was established around the sixteenth century. The soil and climate conditions of this location proved favorable to rose growing, and the number of rose plantations that existed at this time gave this area the nickname "Valley of Roses."

During the rose season (late spring to early summer), the first once-blooming roses such as 'Kazanlik' bloom prolifically for a few weeks. The idea of picking roses all day sounds rather delightful for the rosarian or rose lover; however, rose picking can be understood to be a very laborious process. Timing is essential: to get the highest quality of rose oil, harvesting should occur before the sun rises and evaporates the natural oils. Once picked, the roses are taken to the distillery where the process immediately takes place. In today's operations, steam distillation is a method of low-pressure steam penetrating the fresh blossoms to release the rose's oil into a vapor. The accumulated vapor is then condensed by cooling it through a coil. Finally, the rose oil and water easily separate from each other due to the difference in their density.

Today, the Bulgarian Rose Distillery claims its rose oil and rosewater products are some of the best in the world. But before we place a few orders, let us take a look at the much-trialed process of distilling roses and some of the chemical combinations involved.

CHEMISTRY OF ROSE FRAGRANCE

Bulgarian rose oil owes much of its "true rose" odor to the presence of beta-Damascenone, which is detectable in scarce concentrations. Beta-Damascenone belongs to a family of chemicals known as rose *ketones*—and as an interesting side note, according to Peter Schieberle, (E)-beta-Damascenone is identified as a primary odor of bourbon made in Kentucky. A review of *Rose (Rosa damascena)* by John C. Leffingwell, says that Bulgarian rose oil has more than 275 ingredients: the highest proportion of oil was citronellol at 0.38 (delivering only 4.3 percent of the total odor units). In comparison, beta-Damascenone was 0.0014 (a much smaller proportion, yet providing a whopping 70 percent of odor units, that is a "significant majority of the odor contribution"). The discovery of rose ketones allowed the creation of dramatically new type of perfumes, as exemplified by "Poison" by Christian Dior, 1985, wherein Damascenone and the alpha- and beta-Damascenone may be used at quite high levels.

RIGHT Rose picking in Bulgaria would start early in the morning before the oil evaporated, securing the highest quality.

EARLIER DISTILLATION PROCESSES

WE HAVE ALREADY READ SOME STORIES OF ROSES, rose water and distillation. Aristotle taught his philosophies of distillation where macerating flowers and spices made perfumes within oils and fats—never with water or steam (this came later). Writings like Aristotle's, documenting particular "instructions," allowed the distillation of roses to be put more widely into use.

Ancient Greeks, however, were by no means the first to perfume their oils. In the thirteenth century BCE, Pylos had a flourishing industry making rose and sage-scented oils. The Olive Oil Tablets of Pylos gave details of the number of scented oils being produced and traded (see page 54). Other civilizations at the time, particularly the Persians, loved their lavish ointments and fragrances.

RIGHT Women making rose water from petals from an illustrated medieval handbook of health called *Tacuinum Sanitatis.*

Alexander the Great famously encountered their healing properties after he had defeated King Darius and the Persians at the Battle of Gaugamela, 331 BCE. Plutarch (49–c.119 CE) states in *Greek and Roman Lives* that Alexander was going to the Persian king's tent to "cleanse ourselves from the toils of war in the bath of Darius."

Here, when he beheld the bathing vessels, the water-pots, the pans, and the ointment boxes, all of gold curiously wrought, and smelt the fragrant odors with which the whole place was exquisitely perfumed [with roses], and from thence passed into a pavilion of great size and height, where the couches and tables and preparations for an entertainment were perfectly magnificent, he turned to those about him and said, 'This, it seems, is royalty.'

Greek and Roman Lives PLUTARCH, translated by Dryden and Clough, 2005

The Greeks quickly incorporated the Persian liking of perfumery, as documented by Theophrastus. Most Greek perfumes used a base of Egyptian or Syrian olive oil, pressed from coarse olives. However, for roses, sesame oil was used because of its thicker qualities that held the fragrance. Rose perfumes, at that time, were made by steeping the petals with ginger grass, aspalathus, sweet flag, and salt.

Aspalathus is a spiny plant called "broom" and is native to the Mediterranean climate. Pliny the Elder in his *Naturalis Historia* speaks of aspalathus as a diminished tree or shrub with flowers that resemble a rose. The sweet smell of the plant's root is used in perfumes. Theophrastus states:

This treatment of pressed leaves in oil is peculiar to rose perfume and involves a great deal of waste, twenty-three gallons of salt being but to eight gallons and a half of perfume.

THEOPHRASTUS

He goes on to write that the rose perfumes were one of the first to receive a color—red, from the alkanet. Related to the borage family of plants, alkanet was a source of red dye at the time. (Theophrastus, *Enquiry into Plants*).

Roman Distillation Process

Pliny the Elder assimilated as much information as he could about perfumes into his *Naturalis Historia*. This written attempt reinforces the Roman enthusiasm for their oils, unguents, and the fragrance of roses. He explains that Romans made their unguents from oils and sweet-smelling substances, adding color not only from alkanet but also with the addition of cinnabar. This mineral gave a vermillion color.

Displays in the Houses of Pompeii

We can find some evidence of the Roman distillation process in scenes on the walls of the House of the Vettii in Pompeii. The Vettii brothers grew roses in their garden, manufactured garlands, created perfumes, and even possessed a goldsmith's store. They wanted to depict themselves as successful business owners different from the residents of Ostia, who were mostly working class. A particular panel shows a scene depicting the production of perfume and its trade. Cupids are working a press, a psyche figure is stirring liquid, and other cupids are mixing a concoction in a different container.

RIGHT A fresco from the House of the Vettii in Pompeii shows a vast scene of perfume making. A female figure on the left is experimenting with scent.

This panel illustrates a distillation process of sorts, and the use of divine or mythological figures give credentials to the store owner and his trade.

Next to this scene is another cupid, opening bottles and flasks for the customer identified as Venus. Some scholars depict this scene as the production of pharmaceuticals or wine. However, the left side of the fresco shows a female figure smelling a fragrance and holding her wrist to her nose—this female appears to represent a customer buying (or testing) the perfume. Additionally, other scenes show cupids and psyches collecting flowers and producing garlands.

The House of Calpurnii, also in Pompeii, is described as having displays and frescoes depicting the same scenes. These, unfortunately, have been lost or destroyed. The frescoes of both of these houses depict the production and marketing of fragrances or perfumes. These stores were small and sometimes were integrated into the front parts of the houses of the elite. Other points of retail could be found at the Macellum, a marketplace with temporary stalls and booths. Pliny states: "perfumes were among the most elegant and honorable enjoyments of life."

Pliny also suggests that the next most important site for perfume production during the era—after Alexandria in Egypt—was Campania.

Perfume in Roman, Christian, Arab, and Persian Worlds

The Roman perfume industry had a home in Capua, the chief city of the Campania region, as well as Pompeii. Archaeological evidence such as flasks, bottles, and oil extraction wedges show proof of this. After wool, it is recorded that perfume production (although more costly) was the second most important industry in Pompeii. Fashionable products such as rose perfumes and unguents could only be afforded by the wealthy. However, even those less fortunate would consume perfumes, which were scented with local flowers (including roses). Interestingly, Pliny criticized women for the use of perfumes for attention, describing perfumes as the "most superfluous of all forms of luxury" (Larry Shiner, *Art Scents*).

Paralleling the absence of the pleasure garden after Rome's fall, we also find the lack of the use of roses in perfumes, and perfumes in general. Early Christians hated the use of smells, and once again, roses were associated with pagan worship. Gradually, the Church found the symbolism of the rose in Mary (see Chapter 5), and with the rose, perfumes made their way into Christian rituals.

Leaving Christianity, we find advances in perfumes—mainly of roses—within the Arab world. Of note here is the prophet Muhammad's heartland of Arabia. From the work of the Persian physician, philosopher, and alchemist, Al-Rāzī (known to the Latins as Rhazes), we find reference to the *al-anbîk*, which translates to: "that apparatus in which rose water is made." As technologies changed, distillation became an art.

Persia has long been celebrated as a country of roses. Its perfume industry, as it was, existed at least from the beginning of the ninth century CE. The large rose fields of Shiraz provided the center for rosewater making. The application of perfumery is often attributed to the great eleventh-century Persian physician and polymath, Ibn Sînâ (Avicenna) (see page 157).

The rose water industry was so important to the province of Faristan that a yearly payment of 30,000 vials of rose water to the Treasury of Caliph al-Ma'mun in Baghdad was required as far back as 810 CE (see page 40). From here, rose water was sent to far regions of the world: China, India, Egypt, and North Africa. The German traveler and physician, Engelbert Kaempfer, mentions while traveling to Shiraz in the 1680s (see below):

Vivid descriptions of the area of Damascus paint a picture of a village known as al-Munazzah. The renowned Arab cosmographer and religious leader, al-Dimashqi, who died in 1327, describes this area (known as "the Incomparable"):

Even as the roses in Persia are produced in greater abundance and with finer perfume than those in any other country in the world, so also do those of this particular district in the vicinity of Shiraz, excel in profusion and in fragrance those of any other locality in Persia.

Rhodologia,
JOHN CHARLES SAWER, 1894

Because of the healthy air, the pure water, the beautiful pleasure-houses, the delicious fruit, the many flowers and roses and the production of rose water; the residue of which is thrown on the roads, lanes, and alleys of this place like dirt. Thus, the smell is incomparable and finer than musk until the roses are overblown.

A Short History of the Art of Distillation,
ROBERT J. FORBES, 1970

THE RAGE FOR ROSE OIL AND PERFUME

ROSE OIL IS ALSO A PRODUCT OF DISTILLATION. **An oily layer is left once the rose water has been captured. This rose oil is used to produce "attar" or "otto" of roses in a second distillation. An Italian, Geronimo Rossi of Ravenna (c. 1539), developed the technique to separate this oil from rose water. Rossi himself later described his discovery, and distilled rose oil appears in the price lists of German apothecaries from the 1580s.**

Rose oil was "discovered" by chance in 1612 at a wedding feast given by the beautiful princess Nur Jahan for her husband, the Mughal Emperor Jahangir. The princess spared no expense as she filled an entire canal with rose water. As she walked with the emperor by the edge of the fragrant water, they noticed an oily scum floating on its surface. Once the oil was skimmed off, the whole court recognized it as the most delicate perfume known to the East. It was claimed that (see right):

If one drop is rubbed on the palm, it will perfume a whole room and make it seem more subtly fragrant than if many rosebuds had opened at once. It cheers one up and restores the soul.

The Rose, JENNIFER POTTER, 2010

The trade connection through Constantinople, which stood at the crossroads of the East and West, aids in explaining the Italians' rise in the perfume industry in the 1600s–1700s. From Italy, the taste for refined perfumes (and the skill to make them) spread throughout Europe. In England, at the time, the Earl of Oxford, Edward de Vere (1550–1604), created a craze for sweet waters and perfumes at the court of Queen Elizabeth I. It became custom at this time for the households of the day to have recipes and such for perfuming clothes and masking body smells with rose water and other items. The Italians had become experts in the making of unguents, and other countries were eager to learn.

ABOVE Nur Jahan and her husband, the Mughal Emperor Jahangir, holding delicate pink roses in their hands. A story credits Nur Jahan's mother with discovering the attar of roses.

Europeans of the sixteenth century used perfume to mask unpleasant odors. The garden historian Jennifer Potter writes that the French author of a self-help guide of 1572 advises: "To cure the goatlike stench of armpits, it is useful to press and rub the skin with the compound of roses" (*Seven Flowers and How They Shaped Our World*).

French Perfume

In the eighteenth century, the French became more important to the perfume industry than the Italians. By the 1750s, the French love of perfumes had become all the rage. Scents had left the musk range and entered the sweet smell of flowers. Rose water was at the forefront, along with violets, thyme, lavender, and rosemary. This rose love is demonstrated through King Louis XV's mistress, Madame de Pompadour, who in a portrait by the French artist François Boucher is shown surrounded by roses. This painting illustrates the rococo style of the time and displays the noblewoman with roses (possibly *Rosa centifolia*) strewn among her exuberant gown.

Roses and their perfumes dominated the scene for another century, where we find Empress Joséphine at her rose garden at Malmaison. In the world of roses, Joséphine was a significant influencer of her time. Strangely enough, she is said to have favored musk scents over the florals: "Sixty years after her death, her boudoir at Malmaison still smelt unmistakably of musk." (*The Rose*, Jennifer Potter). After Joséphine's passing, floral scents (including the rose) came back into fashion.

RIGHT *Madame de Pompadour* (1756) by François Boucher shows her famous ornate style with the exuberant use of roses throughout the composition. The two roses on the floor by the dog symbolize the couple, one for the king and one for Madame.

BOTANICAL ART

ANCIENT GREECE AND ROME gave inspiration to the Neoclassicism art and culture of this new century. *Rosa damascena* and other fragrant rose species, like the Centifolias, inspired artists to create masterpieces in still life. Behind the religious and historical paintings, still-life painting was considered an inferior form of art. The Dutch began to lead this type of artistry and the city of Antwerp became a hub. The artists began painting monumental still-life paintings featuring flowers.

Along with Pierre-Joseph Redouté (see Chapter 8) and his ability to capture flowers and roses, Rachel Ruysch (1664–1750) was another excellent artist in the genre, a painter of beautifully detailed works. Her father was a botanist and she started painting insects and flowers in his collection before being apprenticed to a flower painter at the age of just fifteen. She became internationally renowned for her flower paintings and had a long and successful career, alongside having ten children with her husband, Amsterdam portrait painter Juriaen Pool (1666–1745). Her *Still Life of Flowers with a Nosegay of Roses, Marigolds, Larkspur, a Bumblebee and Other Insects* (1695) includes wonderfully vibrant examples of flowers and insects.

Ruysch's paintings of flowers were elevated in such a way that she had no rival in what is known as the Dutch Golden Age of Art. For sixty years of painting, Ruysch created her flower bouquets with signature asymmetrical

Rachel Ruysch's *Roses, Convolvulus, Poppies, and other Flowers in an Urn on a Stone Ledge* (1688).

1. Insects

2. Cupped flower form

3. Rose thorns

views, which became a trademark. These arrangements produced a more realistic and three-dimensional view of the flowers, which brought their realism to a higher level.

Ruysch's painting *Roses, Convolvulus, Poppies, and other Flowers in an Urn on a Stone Ledge* (1688) shows with much accuracy the title's flowers. The white rose interestingly shows a few insects—almost as if they are manipulated by the rose's fragrance and are searching for a feast of pollen (1). The pink rose

that hangs above a milk thistle is a beautiful example of the cupped flower form of the Centifolias (2). The softness of the rose petals is of extreme contrast to the sharpness of the jagged edge of the milk thistle leaf just below. The painting shows so much life in such exquisite detail that the thorns on the rose stems can almost be touched (3), and the fragrance of the rose seems to permeate from this magnificent painting.

Flowers in a Wooden Vessel (1603) by Jan Brueghel
is a large vanitas painting. Gathering flowers
together that would not be blooming at the
same time reminds viewers of their mortality
and to enjoy all the flowers in the present.

Flower Brueghel

Jan Brueghel the Elder (1568–1625), also known as "Flower Brueghel," was one of many working in the genre of flower still life. He painted with realistic, botanically correct rendering, using tulips, irises, and roses to anchor his bouquets. He was one of the first of his class of baroque painters to feature Centifolia roses.

This type of flower collage is unusual, given that many of the flowers depicted in one painting would not be blooming together in the garden. The compositions are flowers from all the seasons brought together in one vase. Brueghel's *Flowers in a Wooden Vessel* (1603) shows one of the biggest gatherings of flowers in art. As you look at the painting, you might ask if the lilies, tulips, fritillaries, daffodils, snowdrops, carnations, cornflowers, peonies, anemones, and roses would have blossomed at the same time? The answer is no; however, this idea became known as vanitas painting (reminding the viewer of their mortality) or *memento mori* (of mortality), combining the real, the ideal, and the symbolic in one vase. Further, at the request of the Italian Cardinal Federico Borromeo (1564–1631), Brueghel became known for his paintings of flower garlands. In *Madonna in a Floral Wreath* (1621), his use of roses and other flowers encapsulates the Virgin Mary.

Amid the riches of the time, Dutch artists created moralizing still-life paintings that reminded viewers of the fleeting nature of material wealth. These paintings often featured skulls to signify death, hourglasses to indicate the passing of time, and wilting flowers to symbolize the temporary. While the vanitas scenes signaled the transient nature of all living things, these bursting bouquets demonstrated the ability of art to freeze time and grant flowers for our present joy in life and the eternal life to follow.

REASONS TO GROW ROSES

These quotes from two of my garden heroes illustrate the importance of growing roses for scent—and I couldn't agree more:

The fragrance of roses has the magical ability to be able to calm you down and, at the same time, raise your spirits.
MICHAEL MARRIOTT, 2015

Through smell, I believe we are getting closer to understanding the mystery of the rose and why this flower, above all others, retains such a powerful hold on our imagination.
The Rose, JENNIFER POTTER, 2010

We can complain because rose bushes have thorns, or rejoice because thorns have roses.

ALPHONSE KARR, 1856

Tea rose 'Hyménée'

8

THE ROMANTIC ROSE

(1790–1850)

The Romantic movement celebrated the primacy of
emotion and the individual, while glorifying the past
and the natural world. The "romantic rose" had complex
symbolism: beauty was troubling because it contained
the seeds of its passing, of decay, and the rose
represented extremes—love and sorrow, joy and grief,
life and death. A major breakthrough came when,
following the importation of roses from China, the
crossing of Chinese with European roses made possible
"remontancy"—flowering more than once in a season.

CHINESE ROSE IMPORTS

Chinese roses (*Rosa chinensis*) were imported to Europe and North America in the 1790s and early 1800s as part of a general enthusiasm for Oriental products and decorative arts, including teas, porcelain, silks, and carpets.

Before their arrival, the gardeners' mainstays were the European 'Old Roses' that they had inherited from the medical and classical gardeners. These—Gallica, Damask, Alba, Centifolia, and Provence types—mainly flowered very beautifully but briefly in early summer (depending on temperature and region), which is why the rose became a symbol of evanescent loveliness, a great beauty that could not last. There were a few exceptions, for example, the musk rose (*Rosa moschata*) could flower from late spring until fall (conditional on the climate). The 'Autumn Damask' has a repeat bloom in the fall. But in the main, roses were limited to summer flowering in Europe. While the Chinese roses typically had less fragrant, smaller blooms, they could flower repeatedly through summer and into late fall. They are also mutable (liable to change) after first blooming, for example, 'Mutabilis' is a Chinese rose that goes from peach to darker pink. Some European roses tended to grow paler after they opened—this was due to many factors, such as climatic conditions.

Therefore, the characteristics of the China roses were highly attractive to gardeners, and they caused a sensation in Europe and North America. Several prominent figures developed an enthusiasm for roses and collected and cultivated the flowers, which became the basis of modern roses. In the United States, President Thomas Jefferson

(1743–1826; president 1801–1809) grew roses on the estate he developed at Monticello, in Virginia.

In France, the Empress Joséphine (1763–1814), consort of Napoléon Bonaparte who ruled as Emperor Napoléon I in 1804–1814 and 1815, bought the Château de Malmaison, a rundown estate of 150 acres (60 hectares), situated 9 miles (15 km) west of the center of Paris. There she spent a vast sum renovating the house and land, building an orangery and a vast greenhouse heated by twelve coal-burning stoves, and planting a widely celebrated rose garden in which she supported rose breeding and grew 250 varieties of rose. She invited the celebrated Belgian botanist and artist Pierre-Joseph Redouté (1759–1840) to paint her roses. He made detailed and beautiful watercolors of roses

and other flowers at Malmaison, and many were published as engravings. Redouté was celebrated as "the Raphael of flowers" (a reference to the Italian Renaissance master Raffaello Sanzio da Urbino); he is considered the greatest botanical artist in history, a worthy successor to the great seventeenth-century Flemish and Dutch flower painters, such as Jan Brueghel the Elder and Rachel Ruysch.

RIGHT *Rosa lutea* 'Sulphurea' and *Rosa* 'Hume's Blush Tea-Scented China' from *Choix des Plus Belles Fleurs* (1827) by Pierre-Joseph Redouté, who was invited by Empress Joséphine of France to paint the 250 rose varieties at the Château de Malmaison.

Four China Studs

The first China rose to arrive was probably the pink rose 'Old Blush,' called *Yue Yue Fen* ('Monthly Pink') in China and known as 'Old Blush China,' 'Old China Monthly,' and 'Parsons' Pink China' in Europe, after a certain Mr. Parsons who introduced it to England in 1793. Two red roses, *Yue Yue Hong* ('Monthly Red') and *Chi Long Han Zhu* ('White Pearl in Red Dragon's Mouth'), later called 'Sanguinea' or 'Miss Willmott's Crimson China,' arrived around the same time.

The pink 'Old Blush' and the 'Crimson China' roses are usually included in a list of four China studs introduced at the time. The other two are 'Hume's Blush Tea-Scented China' (1809) and 'Parks' Yellow Tea-Scented China' (1824). These were so-called because their fragrance resembled the scent of the China teas drunk by the upper classes in England during this period. The 'Hume's Blush Tea-Scented China' was sent from China by tea inspector John Reeves (1774–1856) to Lady Amelia Hume (1751–1809),

a British horticulturalist who collected and grew plants at her estate in Wormleybury, Hertfordshire; both Lady Hume and her husband Sir Abraham Hume, 2nd Baronet (1749–1838), were keen gardeners and imported exotic plants. When the rose was displayed in Chelsea, London, it was much in demand and, remarkably, a special truce in the war between Britain and France was arranged by Emperor Napoléon I so that Empress Joséphine could safely order and receive samples for her collection at the Château de Malmaison.

'Parks' Yellow Tea-Scented China' was brought to the Royal Horticultural Society from China by a gardener, John Damper Parks (c. 1791–1866), whom the Society had sent specially for the purpose. It was called *Danhuang Xianshui* ('Light Yellow Sweet Water'); its petals were pale yellow and turned ivory white after opening. Parks returned with a remarkable consignment, including the first aspidistra to be brought to Europe as well as sixteen chrysanthemums and the rose.

RIGHT A Chinese watercolor of the pink rose 'Parsons' Pink China,' which came to Europe from China. It was recorded in Sweden in 1752 before Parsons brought it to England.

A 'Blush Noisette' rose painted by Pierre-Joseph Redouté;
this rose was created by crossing 'Champneys' Pink Cluster'
with an unknown seedling to produce a light-pink rose in
around 1815. It is also known as 'Noisette Carnée'
(Flesh-coloured Noisette).

Rose Breeding

In the United States, John Champneys (1743–1820) of Charleston, South Carolina, introduced the first hybridized rose to North America in 1811, *Rosa moschata* × 'Parsons' Pink China' to produce what was called 'Champneys' Pink Cluster.' He gave the rose to a nurseryman named Philippe Noisette, who sent seedlings and seeds to his brother Louis (1772–1849) in France. There, French enthusiasts crossed this 'Pink Cluster' with the 'Parks' Yellow Tea-Scented China' to produce a new class of roses called the Noisettes.

By the mid-nineteenth century, rose breeding was very well established. A rosarium planted at Abney Park Cemetery in London, U.K., contained more than a thousand rose species and cultivars.

Around the same time, one of the first American rose nurseries was set up in Philadelphia, by Scottish-born botanist Robert Buist (1805–1880) who, after training at the Edinburgh Botanic Gardens, had come to the United States in 1828 and worked at Lemon Hill, the mansion at Fairmount Park, Philadelphia, owned by a merchant, Henry C. Pratt, who had one of the finest gardens in the United States at the time. Buist went on to write the first rose-growing guide, *The Rose Manual*, in 1844.

REMONTANCY AND THE ROSE

The introduction of Chinese roses to the West made a major breakthrough possible—the breeding of the first roses outside Asia to flower more than once in a season. They were called remontant—from the French verb *remonter* ("to climb up again"); their counterparts that flower only once were called non-remontant, once-flowering, or summer-flowering roses.

Few wild roses are remontant—some exceptions are the China rose (*Rosa chinensis*) and *Rosa rugosa*. Remontant roses were cultivated in China and other parts of Asia from around 1000 CE. Their crossing with European roses in the Romantic period made the growing of cultivated remontant roses possible anywhere in the world.

ROSE ODDITIES

THERE ARE NO BLUE ROSES IN NATURE. **Rose breeders have developed "blue roses" by hybridization, but the color is really lilac. In 2004 an Australian company named Florigene and a Japanese company, Suntory, developed a blue rose, 'Applause,' by genetically engineering of a white rose: they inserted a gene for the blue plant pigment delphinidin from a pansy into a red 'Cardinal de Richelieu' rose. They then tried to stop production of the red coloring in the hope of creating a blue flower. But this complex genetic operation failed to block the red coloring completely, so the final flower combined red and blue and ended up being more a mauve or lilac color than a true blue.**

The Legend of the Blue Rose

Because it is impossible to find in nature, the blue rose is often used as a symbol of impossible or secret love. In the Chinese legend of the blue rose, an emperor declared that he wanted to find a husband for his beautiful, deeply intelligent daughter. There were many suitors: but the princess came up with a subtle idea to block her father's plan and declared that she would marry the suitor who was able to find and bring her a blue rose. Most suitors saw that this demand was not possible and melted away, but three persisted in the hope that they could achieve the impossible: a merchant, a warrior, and a chief justice.

The warrior went to a distant kingdom where the local monarch provided a beautiful blue sapphire carved in the shape of a rose. But when he returned with it, the princess said that—beautiful as it was—it was not a real rose and so she turned him down. The merchant decided that money was no object and went to the best florist and offered him a fortune to find a blue rose. The florist's wife took a white rose and painted it blue, and the merchant took this to the princess. She appreciated the ingenuity, but turned down the merchant because this was not truly a blue rose. The chief justice gave the princess a beautiful glass with a blue rose painted on it, but the princess rejected this final attempt.

The princess seemed set for an independent life without a husband. But then a honey-voiced troubadour visited the palace and sang her the most beautiful songs and the two fell in love. The troubadour came to the princess and presented her with a fine white rose that he had just picked in the palace garden and she said that this was the blue rose she had been looking for. Her father the emperor and the people were amazed, but all had to admit that since the princess had set the condition of receiving a blue rose, and since she now said that this rose was blue, there was nothing to be done but allow her to wed the troubadour.

DYEING ROSES

There was a tradition of dyeing white roses blue via their roots to produce an azure blue flower, according to a twelfth-century Arabic tome, *Kitab al-filaha*, by the Muslim Arab author Ibn al-Awwam (d. 1158), based in Seville, Spain. But, not all "blue" roses are dyed. Shown here is the blue-pigmented rose, 'Applause'.

Blue rose 'Applause'

Black Roses of Turkey

Some people claim black roses exist, but they are, in fact, a very dark red in color. False reports on the internet from c. 2008 onward claimed that black roses—called *kara gül* in Turkish—grew in the small town of Halfeti in southern Turkey and the place became famous on the back of the claim. While these flowers appear black—or close to black—when in bud, they are a deep-red wine color when in bloom. This "black rose" is a local variant of *Rosa odorata*, native to China.

Green Roses in History

Unlike their imaginary blue and black cousins, the 'Green Rose' really does exist—and has a fascinating piece of history allegedly attached to it. These roses are certainly odd-looking; a sport form of the 'Old Blush' rose (*Rosa chinensis viridiflora*), in which the petals are in fact the green sepals that in normal roses enclose a bud as it develops. Normally a flower develops first sepals, then petals, then the male reproductive stamens and female reproductive pistils and carpels; but the 'Green Rose' does not progress beyond the sepal-forming stage. The flower head is an odd-looking collection of sepals; and because no stamens, pistils, and carpels are formed, the 'Green Rose' is sterile. It has a peppery fragrance that is not too strong and grows to 2–5 feet (0.6–1.5 meters) tall.

These roses perhaps show us what flowers looked like early in their evolution before they developed brightly colored petals, and they may also have an interesting historical significance, according to Stephen Scanniello, President of the Heritage Rose Foundation. In c. 1800–1850 the Underground Railroad—a network of safe houses and secure routes—was established to help escaped African American slaves flee from southern states to safety in the north and Canada. Some people believe that Quakers, who supported the Underground Railroad, would plant the 'Green Rose' in the front garden of houses that were safe for the escaped slaves to stop in; so a fugitive, seeing the green flowers, would know it was safe to seek food and refuge.

LEFT The 'Green Rose' is a mutated form of 'Old Blush' that was then named *viridiflora*, meaning green-flowered. The "petals" are the green sepals that normally enclose a rosebud.

STILL LIFE WITH ROSES

ROSES REMAINED A FAVORITE SUBJECT for painters of still-life canvases in the Romantic period and through to the end of the nineteenth century. The great Dutch artist Vincent van Gogh (1853–1890) painted *Still Life: Vase with Pink Roses, Vase of Roses,* and another rose still life called simply *Roses* in 1890. He painted these works just before leaving the asylum in Saint-Rémy-de-Provence, southern France, where he had stayed for a year from May 1889.

By painting flowers, some critics say that van Gogh was expressing the optimism he felt about leaving the asylum and moving to Auvers-sur-Oise, on the outskirts of Paris. He loved to paint flowers for their colors—once telling his sister Willemina in a letter about an earlier period (see right).

Around the same time, French Impressionist artist Pierre-Auguste Renoir (1841–1919) painted masterful still-life subjects including *Bouquet of Roses* in c. 1890–1900 and *Roses* (1912).

> ...*I painted almost nothing but flowers so I could get used to colors other than gray—pink, soft or bright green, light blue, violet, yellow, glorious red.*
>
> VINCENT VAN GOGH, 1889

In his later years, he painted roses— especially red roses—over and over again, preferring them to other flowers.

French artist Henri Fantin-Latour (1836–1904) was a celebrated painter of floral still-life subjects, and portraits of writers and artists. *Roses* (1894), *Roses and Lilies* (1888), and *White Roses* (1875) are excellent examples of his work. Fantin-Latour was mentioned by Marcel Proust (1871–1922) in his epic novel *À la recherche du temps perdu* (*In Search of Lost Time*, 1913–1927), in the third volume, *Le côté de Guermantes* (*The Guermantes Way*): "he asked her whether she had seen the flower painting by Fantin-Latour which had recently been exhibited."

THE ROSE AND THE ROMANTIC POETS

THROUGHOUT THIS PERIOD, the rose remained a symbol of youth, freshness, and beauty. Scottish national poet Robert Burns (1759–1796) used the rose in this way in his celebrated 1794 version of the traditional Scots song, *My Love is Like a Red, Red Rose*. In Burns's words:

O my Luve's like a red, red rose,
That's newly sprung in June.
O my Luve's like the melodie
That's sweetly play'd in tune.

A Red, Red Rose, ROBERT BURNS, 1794

Burns's song was based on traditional sources—he gave it to Italian-born singer Pietro Urbani (1749–1816), who published it in the book *A Selection of Scots Songs* (c. 1792–94). In this period, Burns was collecting and preserving traditional Scots folk songs for posterity for the songbook *The Scots Musical Museum*, published by music seller James Johnson, and it seems that Burns heard *A Red, Red Rose* while out in the country and wrote it down. In 1794, Pietro Urbani wrote that the words of *A Red, Red Rose* were given by "a celebrated Scots poet" (i.e. Burns), "who was so struck by them when sung by a country girl that he wrote them down."

But in the English Romantic poet William Blake's (1757–1827) poem, *The Sick Rose*, the rose in the poem is inwardly decaying because a worm is eating at its heart. Blake's verse is reproduced on the facing page.

O Rose thou art sick.
The invisible worm,
That flies in the night
In the howling storm:

Has found out thy bed
Of crimson joy:
And his dark secret love
Does thy life destroy.

The poem was first published in 1794 in Blake's beautifully illustrated book *Songs of Experience*. Some readers see the poem as being about innocence corrupted by experience, or beauty and purity undermined by ugliness and evil. In such a reading, the rose remains a symbol of love, of youthful beauty, of freshness, of innocence; but in a world that is never perfect, the seemingly perfect object has hidden flaws—here the worm that you cannot see. The image seems to resonate with the Christian idea of the fallen world— perhaps the worm suggests the biblical serpent (the devil) who in the Book of Genesis tempted Eve in the Garden of Eden, and as a result Adam and Eve (and their descendants) were condemned to live in a fallen world where there is no perfection. Other readers see the worm as representing death (since worms feed on dead bodies) and the rose therefore might be a symbol of life—for all its beauty and youthful vibrancy, it will, like all living things, die.

Lovely is the Rose

"Lovely is the rose" wrote fellow Romantic poet William Wordsworth (1770–1850) in his poem *Ode: Intimations of Immortality from Recollections of Early Childhood* (composed c. 1804), using the rose again as a symbol of purest natural beauty. In the poem he describes how in childhood the natural world seems full of divine glory, but for a grown man, while the world is still beautiful, this sense of wonder is lost—perhaps while children believe they are immortal, adults know they must die—but he comes in the end to the understanding that knowing about death makes the natural world all the more movingly beautiful, because even looking at a simple flower in the wind gives rise to "thoughts that do often lie too deep for tears." The rose and other beautiful flowers symbolize joy and beauty—but as in Blake's poem—also remind the onlooker of death.

Percy Bysshe Shelley (1792–1822), another of the great English Romantic poets, imagined rose petals as the bedding on which a dear dead friend or lover lay in his elegiac poem of 1821 *Music when Soft Voices Die (To —)*: "Rose leaves, when the rose is dead, Are heaped for the belovèd's bed." As with Blake and Wordsworth, thinking of the rose that blooms beautifully, makes an onlooker consider how beautiful things cannot last and roses must go over—and of death. In Shelley's poem, the rose leaves are an enduring reminder of past beauty and are likened to one's memories of a dead person and their thoughts; the love the deceased person inspired endures as the leaves survive the death of the flower.

THE GARDEN'S PRIDE

For Wordsworth's great friend and fellow-poet Samuel Taylor Coleridge (1772–1834) in his poem *The Rose*, the rose ("the Garden's Pride") was home to love—Cupid, the god of love, here personified as "Love":

> *Within the petals of the Rose*
> *A sleeping Love I 'spied.*

He plucks the rose and pins it on the chest of his beloved, a young woman named Sara:

> *And plac'd him, cag'd within the flower*
> *On spotless Sara's breast.*

The conceit of the poem is that when Cupid awakes and sees the beauty of the poet's beloved, he declares that he will make his home there—giving up the chance to visit Venus herself, so beautiful is Sara. When he first woke, he "stamp'd his faery feet," but then he saw Sara and was calmed:

> *Ah! soon the soul-entrancing sight*
> *Subdued th' impatient boy!*
> *He gaz'd! he thrill'd with deep delight!*
> *Then clapp'd his wings for joy.*

> *"And O!" he cried—"Of magic kind*
> *What charms this Throne endear!*
> *Some other Love let Venus find*
> *I'll fix my empire here."*

THE ROSE AND THE ROMANTIC POETS

Another English Romantic poet, John Keats (1795–1821), dwelt on the different types of beauty found in wild country roses and their cultivated garden counterparts in his beautiful celebration of friendship, *To a Friend Who Sent Me Some Roses* (published 1817). In the poem the poet is walking in the country and came across a beautiful musk rose ("the sweetest flower wild nature yields") and likened it to a wand carried by Titania, the queen of the faeries, before he "feasted on its fragrancy," and he thought it more beautiful than cultivated garden roses: "I thought the garden-rose it far excell'd." But the roses he was sent by a friend, although they were garden roses, were the most beautiful because they carried the associations of their friendship: "Soft voices had they, that with tender plea / Whisper'd of peace, and truth, and friendliness unquell'd."

The Last Rose of Summer

Irish poet Thomas Moore (1779–1852) was also inspired by the passing of the rose's beauty to consider the shortness of life and friendships and of his own eventual passing in the celebrated poem *The Last Rose of Summer*. Moore wrote this in 1805 while staying at Jenkinstown Park, then an estate in County Kilkenny. Reputedly he was inspired by an example of the 'Old Blush,' the first China rose to be successfully established in Europe and North America. The poet admires the last rose of summer, which is standing in the flowerbed after its counterparts have died and decides that rather than leave it standing alone, he will scatters its petals over the ground and will send the flower to rest with former "mates" (see below).

I'll not leave thee, thou lone one,
To pine on the stem;
Since the lovely are sleeping,
Go, sleep thou with them;
Thus kindly I scatter
Thy leaves o'er the bed,
Where thy mates of the garden
Lie scentless and dead.

The Last Rose of Summer,
THOMAS MOORE, 1805

RIGHT The pink 'Old Blush' rose arrived in Europe from China in c. 1750 and became widely popular. A specimen at Jenkinstown Park, Kilkenny, reputedly inspired Thomas Moore to write *The Last Rose of Summer*.

He then considers how he will follow: "When friendships decay, / And from love's shining circle / The gems drop away!", and concludes by deciding that beauty and life must pass, for who would want to be left behind, alone, after the joys have faded—in the way the last rose of summer had been before she scattered its petals? He says: "Oh! who would inhabit / This bleak world alone?"

The poem was set to a traditional tune (*Aislean an Oigfear*, "The Young Man's Dream") and published with music in Moore's *A Selection of Irish Melodies*. The music inspired composers from Beethoven and Mendelssohn to Britten and Hindemith; the song was later performed by artists including Bing Crosby, Nina Simone, and Tom Waits; the poem was later mentioned by major writers including Jules Verne and James Joyce.

THE ROSE IN FOLK TALES

THE MEANING OF ROSES in folk tradition certainly fed into Romantic poetry and literature of the Romantic period. In these years, which saw a movement from country life to work in mills and factories, there was an interest in collecting folk tales to preserve them for posterity—much as Pietro Urbani and Robert Burns had set out to save folk songs for later generations to enjoy.

Brothers Grimm

The German folklorists "the Brothers Grimm", Jacob Ludwig (1785–1863) and Wilhelm Carl (1786–1859) Grimm, published *Grimms' Fairy Tales*, their celebrated collection of German and European folk tales in 1812–1815. Their version of *Sleeping Beauty* was called *Dornröschen* ("Little Briar Rose"), the name of the princess in the tale. During the one hundred years of her sleep, the castle in which she slumbers is partly protected by a magical growth of thorny bushes and trees—in some versions these are roses.

In another of the Brothers Grimm's folk tales, *The Rose*, while the flower embodied beauty it also symbolized its transience; the passing of time and the coming of death. In this troubling tale, a mother sends her son (in some versions it's a daughter) into the forest to gather wood. When the son brings a rose to his mother and says it has been given to him by a boy in the forest, the mother puts the rose into water. When it blooms, a terrible thing happens— the mother goes to wake her son and finds him dead in his bed.

The rose seems again to symbolize beauty and its passing and untimely death in the English fairy tale *The Rose-Tree*, collected by the folklorist Joseph Jacobs (1854–1916) in his *English Fairy Tales*, in which a wicked stepmother kills her husband's beautiful daughter and serves the girl's liver and heart to the father. The girl's brother buries the remains of his sister's body beneath a rose tree, which he waters with his tears and which in time flowers and gives issue to an extraordinary white bird. This bird flies off and returns with a millstone. Three times it bangs the millstone against the eaves of the family house and the third time, the bird kills the stepmother by dropping it on her head.

Hans Christian Andersen

In the tale *Little Ida's Flowers* (1835) by Danish author Hans Christian Andersen (1805–1875), the roses are the king and queen of flowers. The tale tells of how the flowers of the garden attend wonderful balls at night in the palace and palace gardens, and during these events the finest roses occupy the throne as king and queen—and later in the story the flowers hold a ball by night in Ida's own house as she watches, spellbound, from the doorway and sees the roses attend wearing the crowns of the king and queen of the flowers.

LEFT "Queen and King Rose," an illustration from *Little Ida's Flowers*, the fairy tale written by Hans Christian Andersen.

The rose in Andersen's *The Rose Elf* is again a symbol of love and beauty. But it also has a threatening side; the rose and other flowers here have a capacity to reveal secrets and avenge crimes. There is a pantheistic note, too: God dwells in the roses and other flowers and through them can uncover wrongdoing and exact revenge. The tale focuses on an elfin spirit who lives in the petals of a rose flower and becomes involved in the saga of a princess whose lover is killed by her brother; the princess dies of a broken heart before the murdering brother is exposed and himself killed. The murderer is killed by flower spirits living in the petals of a jasmine plant that had grown from the head of the murdered lover. At the end of the tale, the queen bee sings of the power of flowers to avenge evil and how even in the smallest flower is God, who knows all secrets and wrongdoing.

In *The Wild Swans*, roses represent the beauty of the princess Eliza, who is banished by her wicked stepmother, along with her eleven brothers who had been turned into swans. At the story's end she is saved from burning as a witch, as the wood for the fire turns into a rose bush and the air fills with the scent of the most wonderful roses. At its top a single white rose appears that is plucked and worn by the princess as she is married to the local king.

The Rose of Love

In *The Loveliest Rose in the World*, the flower is identified with beauty and love—and in the end with the love of God. A queen loves flowers and roses above all, and a wonderful variety grow in her palace gardens. She falls sick and while most of her doctors say she will die, one says that the world's most beautiful rose can save her. Her people bring roses from gardens and fields, but none revives her; then others think of other types of blooming rose but again none revives her. Finally, her son comes into the room carrying the Bible and they share the story of Christ's death. This brings color back to her cheeks— the loveliest rose seems to rise from the book, the rose that sprang from the blood Christ shed on the cross.

Some folk traditions hold that a rose bush sprung from the place where Christ's blood fell from the cross—after his hands were nailed to the wood and after he was pierced in the side with a lance. The rose is associated with God's love and the beauty of His sacrifice. Other reports tell of how the scent of roses hanging in the air was suggestive of holiness—perhaps of the hidden presence of angels; some people report smelling the scent of roses while they are praying. In the Eastern Orthodox and Byzantine Catholic traditions of Christianity, the archangel Barachiel is depicted holding a white rose and sometimes scattering rose petals.

El Arcángel Baraquiel esparciendo flores by
Bartolomé Román (c. 1600) depicts the
archangel Barachiel scattering rose
petals that symbolize the blessings
of God and the righteous life.

It is the time you have wasted for your rose that makes your rose so important.

The Little Prince, ANTOINE DE SAINT-EXUPÉRY, 1943

9

THE ROSE IN MODERN CULTURE

(1851–)

The rose has been important as a symbol and decorative object in religion, poetry, art, literature, music, medicine, fashion, perfume, home decoration, cuisine, and of course gardens over many centuries. In the modern age, this most beautiful of flowers has also found new forms of cultural expression. We have used, celebrated, and engaged with the rose in new ways through modern literature, and in pop culture—in cartoons, movies, and pop music—as well as symbolically in public events and politics. And we still celebrate the rose through fashion, tattoos, paintings, and sculptures.

Rosa carolina

ROSES IN PUBLIC LIFE

On November 20, 1986 US President Ronald Reagan (president 1981–1989) in his Proclamation 5574 designated the rose as the national floral emblem of the United States. He listed the many meanings the rose had for the twentieth century, and said:

> *More often than any other flower, we hold the rose dear as the symbol of life and love and devotion, of beauty and eternity.*
>
> Ronald Reagan, 1986

Reagan recalled how the first US president, George Washington (president 1789–1797), bred roses; that the rose grows in all the states of the union; and how the official presidential residence, the White House, has its own splendid rose garden. Roses, he declared, represent "love and devotion" and many types of love—the love between men and women, between humankind and God, and the love of one's country. He noted the rose's importance in art, music, and literature and how "we decorate our celebrations and parades with roses… we lavish them on our altars, our civil shrines, and the final resting places of our honored dead." He went on to say: "Americans who would speak the language of the heart do so with a rose."

Roses play a prominent role in public ceremonies. Held at New Year, the Rose Parade on Colorado Boulevard, Pasadena, is the US's largest floral parade. The event was launched in 1890 as a celebration of California's mild midwinter, and in 1902 the yearly Rose Bowl college football game was added to help fund the celebration. The 2020 Rose Parade had 40 floats and 20 marching bands.

The second-largest US floral parade is the Portland Rose Festival, held each June in Portland, Oregon, known as the City of Roses (or Rose City). The city held its first annual rose show in 1889 and is the location of the International Rose Test Garden, founded in 1917 with Jesse Currey (president of Portland's Rose Society) as its first curator. It was intended to be a safe haven for rose hybrids grown in Europe and then those at risk because of World War I. To this day, new rose cultivars are sent to the garden from all over the world. The garden in Portland's Washington Park holds more than 10,000 rose bushes and 650 varieties.

Another "City of Roses" is Kutno, Poland, where a rose festival featuring exhibitions of roses and local folk music has been held annually since 1975. Rose breeders from the local area, around the city of Lodz, exhibit their flowers and the festival attracts florists and experts in floral arrangement from Latvia, Lithuania, and Russia. Other rose festivals include the Festival of Roses at Kalaat M'Gouna in Morocco.

RIGHT The Tournament of Roses Parade has been held most years in Pasadena since its foundation by the local Valley Hunt Club in 1890.

THE LITERARY ROSE

In modern and postmodern literature, the rose has continued to resonate as a symbol, carrying many of the meanings it has held throughout the centuries. In the postscript to his celebrated 1980 novel, *The Name of the Rose,* author Umberto Eco (1932–2016) suggested that the rose has so much symbolic resonance that it is almost neutral—as if it were so pregnant with meaning that it became empty of significance.

In a postscript to *The Name of the Rose,* Eco wrote: "the rose is a symbolic figure so full of meaning that it now has hardly any meaning left." This was part of a discussion of why he chose this name for the book. He wrote that he wanted a "totally neutral title." Set in a fourteenth-century monastery in northern Italy, *The Name of the Rose* is in part a murder mystery but is also rich in ideas drawn from theology, religious history, philosophy, and literary theory. It is full of schema and patterns, with references to the biblical Book of Revelation and earlier authors including Sir Arthur Conan Doyle (1859–1930), creator of the Sherlock Holmes novels and stories; the Argentinian novelist Jorge Luis Borges (1899–1986); French novelist Alexandre Dumas (1802–70); British novelist and short story writer Rudyard Kipling (1865–1936); and others. There is also discussion of philosophy, including the work of Austrian-British philosopher Ludwig Wittgenstein (1889–1951), and a key plot device revolves around a lost book by the Ancient Greek philosopher

Aristotle. The "neutral title" that Eco wanted is appropriate because this book, part of the postmodern movement in literature, refuses finality and meaning—after hours and hours of deductive work, its monk-detective William of Baskerville concludes that there "was no pattern."

In addition to being "neutral," the title may also be a reference to the problem of universals in metaphysics and philosophy, as stated by English Franciscan friar and philosopher William of Ockham (c. 1285–c. 1347):

"There is no such thing as a universal rose, only the name rose." The last line of Eco's book reads: "the rose of old remains only in its name; we possess only naked names."

In his postscript, Eco also refers to this poem by a seventeenth-century Mexican mystic nun and poet Sor Juana Inés de la Cruz:

Red rose growing in the meadow,
  you vaunt yourself bravely
bathed in crimson and carmine:
  a rich and fragrant show.
But no: Being fair,
You will be unhappy soon.

SOR JUANA INÉS DE LA CRUZ (1651–1695)

In this reference, the symbolism of the rose is much more familiar: a beautiful living thing that, being full of life and beauty, will not endure.

The Rose as Muse

The rose plays a central role in the profound and remarkable novella, *The Little Prince*, by French aviator and author Antoine de Saint-Exupéry (1900–1944), which was published in both French and English in the United States in 1943. It represents—as is traditional—beauty and love, here with a personal focus, since the rose in the story is believed to represent one particular woman, the author's wife. This intriguing narrative, cast as a children's book, tells the story of the aviator-narrator's plane crash in the Sahara Desert and his encounter there with a laughing little prince with golden hair. The prince is a space traveler and the rose in question is growing on his home planet, a small asteroid called B 612, which also features three volcanoes, one of them extinct. The prince describes how he fell in love with this rose and looked after her attentively, making a glass globe to shelter her from the wind, but he also felt the flower was a little vain and given to exaggeration of her problems, so in the end he left her to explore the universe. On their parting she was serious and expressed regrets that she had not shown him how much she loved him; she said she did not need the globe and would look after herself—and later the prince was very sorry that he had not understood his rose and how to love her. Later, after his landing on Earth, the prince encountered an expanse of rose bushes and was shocked because he had thought his rose was unique.

RIGHT An illustration from *The Little Prince* by Antoine de Saint-Exupéry. In the Sahara Desert the airman-narrator crashes and meets a little boy with golden hair, a lover of roses who has traveled to Earth from a distant asteroid.

Commentators—among them Saint-Exupéry's biographer Paul Webster—have described the rose as a representation of the author's wife, Consuelo de Saint-Exupéry. Antoine and Consuelo had a difficult marriage but, according to Webster, Consuelo was Saint-Exupéry's muse to whom he "poured out his soul in copious letters." She was from El Salvador, a small country known as "the land of volcanoes" and so a clear model for the prince's volcano-troubled home planet in which the beautiful rose was growing. (Consuelo wrote a memoir of their life together entitled *The Tale of the Rose*, which was published posthumously in 2000.) The roses the prince encountered in the desert seemingly represented the author's doubts about his marriage—and possibly a husband's sexual infidelity—since they suggest the wider availability of a plant the prince had once believed was exclusive to his world.

Rose Symbolism in Short Stories

Another modern classic with intriguing rose symbolism is the short story "The Possibility of Evil" by American horror and mystery writer, Shirley Jackson (1916–1965), published in the *Saturday Evening Post* in 1965. It tells of an apparently harmless old lady, Miss Strangeworth, who is very proud of the pink, white, and red roses in her rose garden and her orderly house, but who, it emerges, is the author of a series of mean, anonymous letters sent to people in the small town in which she lives. The letters spread secrets and gossip abroad and are an expression of her feeling that the world is really sick and bad. One of her letters is dropped and a boy delivers it to the person to whom it is addressed. The next day, Miss Strangeworth receives a letter of the kind she sends to other people: this letter writer makes plain that her prized roses have met with a terrible end.

The roses were planted by Miss Strangeworth's grandmother, so they represent her connection to the town in which she lived, and she is unwilling to let people pick them. To her mind, "The roses belonged on Pleasant Street and it bothered [her] to think of people wanting to carry them away." The roses appear to represent her desire for perfection and vision of an idealized life. Since roses combine beauty with thorns, the flowers here also represent Miss Strangeworth's outward pleasantness but inner unpleasantness.

In a short story by William Faulkner (1897–1962), "A Rose for Miss Emily," the rose has connotations of secrecy and, perhaps, preserved beauty. Faulkner's story was published in *The Forum* in 1930. As we saw in earlier chapters, the rose has had connections with secrecy and silence from the Ancient Egyptian and classical period onward. The story is that of a reclusive, elderly Southern woman named Emily Grierson and her relationship with her butler, Tobe, and a working man named Homer who disappears. After her death, people find the decomposed corpse of Homer on her bed and a lock of her hair on the pillow beside it, indicating that she had been sleeping with his corpse for years. The rose might therefore symbolize preserved beauty—Homer's body was kept like a pressed rose between the pages of a book. It might indicate a hidden thing—she secretly kept Homer's body after he died. Roses also symbolize love and, here, the love of Emily for Homer.

LEFT Roses are a symbol of perfection, of love, and of beauty. Dying roses suggest that such ideals cannot last no matter how hard we try to defy death and time, themes expressed in the works by Shirley Jackson and William Faulkner.

THE SCREEN ROSE

THE TWENTIETH CENTURY delivered the visually powerfuland emotionally impactful new art form of cinema, and film directors have found many roles for the rose. As in folk and fairy tales, in literature and fine art, and in religious traditions, in the movies the rose is a symbol of love and the beauty that must pass, of the sadness that accompanies the passing of time, and of mystery—given the way the rosebud unfolds and reveals its secrets.

A red rose is a symbol of love and beauty.

Beauty and the Beast

The rose plays a key role and carries much of this symbolic meaning in the celebrated 1991 Disney animated film, *Beauty and the Beast.* The movie was based on a fairy story of 1740 by French author Gabrielle-Suzanne Barbot de Villeneuve (1685–1755), as reworked by Jeanne-Marie LePrince de Beaumont (1711–1780), and in part on the 1946 movie of the same name directed by Jean Cocteau (1889–1963).

In the Disney movie, an enchantress asks for shelter at a French castle and offers a rose in payment to the local prince, but he is arrogant and lacking in compassion and refuses her request. She casts a spell, turning the prince into a beast and his servants into household objects, and declares that he will remain a beast forever unless he learns to love and is able to win the heart of a young woman before the enchanted rose's last petal falls—in his twenty-first year. A beautiful young woman, Belle, visits the castle and the pair meet. Before the last petal falls, the Beast dies in Belle's arms after being attacked by a local hunter; but because she expresses her love for the Beast the spell is broken, the Beast revives, and becomes the handsome young man he once was. The pair live happily ever after.

The animated film was remade as a live-action picture in 2017, directed by Bill Condon and starring Emma Watson as Beauty and Dan Stevens as the Beast. The flower here seems to symbolize beauty, love, the passing of time, and also mystery—since, as often in fairy tales, appearances are deceptive and people are not who they seem. (In the original fairy story Beauty asked her father, a merchant fallen on hard times, for a rose and he picked one at the Beast's castle when he stopped there for shelter; the Beast threatened to imprison the merchant but allowed him to return home to give Beauty the flower on condition that one of the merchant's daughters agreed to take his place in captivity; Beauty took his place and eventually fell in love with the Beast, who at last revealed his true identity as a prince.)

RIGHT A French-language poster for the 2017 movie version of *Beauty and the Beast*. The Beast must prove he can win the heart of a young woman before the enchanted rose drops its last petal in his twenty-first year.

American Beauty

Red roses starred also in the Academy Award-winning 1999 movie *American Beauty*, directed by Sam Mendes. It stars Kevin Spacey as a magazine executive named Lester Burnham who has a mid-life crisis incorporating troubling sexual fantasies about his daughter's best friend. Red rose petals play a large part in these fantasies; in addition, the house he shares with his wife Carolyn is full of roses in vases. The rose seems to symbolize Lester's lust and selfish desire to be free from his responsibilities, but also suburban conformity—the opposite of freedom. The vivid red of the roses is mirrored in the blood he sheds when he gets his comeuppance at the movie's climax.

V for Vendetta

In the striking 2005 thriller, *V for Vendetta*, roses represent freedom and happiness, as well as the fragility of beauty in a harsh and functional world. The movie, directed by James McTeigue, is based on a 1988 graphic novel by Alan Moore (illustrated by David

LEFT The rose used in the movie *American Beauty* is called 'American Beauty,' a deep-pink cultivar bred under the name 'Madame Ferdinand Jamin' in 1875 in France by Henri Lédéchaux.

Lloyd), and depicts a dystopian future in 2032 in which an anarchist freedom fighter-vigilante named V (played by Hugo Weaving) combats Britain's fascist government. Roses were believed to be extinct, but V is able to cultivate them and leaves a single scarlet rose beside the body of each of the people he kills, who all once tortured him at Larkhill Resettlement Camp.

V grows the roses in memory of his friend Valerie Page, a lesbian tortured and finally killed on account of her sexual orientation. In a letter she recounts how her partner grew scarlet roses for her when they were happy in their London flat—"our place always smelled of roses" during "the best years of my life"; and she writes: "for three years I had roses and apologized to no-one."

The roses seem to represent the joy of that time and the women's freedom to be who they wanted. In the movie, the red 'Grand Prix' roses used in shot are given a made-up name: 'Scarlet Carson'; but in the source novel the roses in question were *Rosa* 'Violet Carson,' a real rose cultivar developed by celebrated Northern Ireland-born rose hybridizer Samuel McGredy IV (1932–2019) and named in tribute to the English actress Violet Carson.

Rosemary's Baby

Roses are troublingly linked to violence against women in the 1968 horror movie *Rosemary's Baby*, directed by Roman Polanski and starring Mia Farrow as a young woman. Rosemary's New York apartment is next to one used by a satanic coven, and while drugged, she is raped and becomes pregnant with the devil's son. Rosemary is shown handling roses and her husband gives her roses. The flowers seem to represent Rosemary's purity—perhaps her vulnerability—and the female reproductive energy that the Satanists are hoping to harness to their own wicked ends.

Citizen Kane

In Orson Welles's classic movie, *Citizen Kane* (1941), the word "rosebud" plays a key role in the drive of the narrative—although the flower itself is not involved. The main character, newspaper magnate Charles Foster Kane (played by Welles himself), utters the word "rosebud" on his deathbed and a reporter is set the task of discovering its meaning. The movie depicts Kane's life but the reporter is unable to find the answer; a final scene indicates that Rosebud was the brand of his childhood sled—and the word represents Kane's memory, looking back on his deathbed across a long life, of a time when he was innocently happy.

The Rose in Movie Soundtracks

The 1947 movie, *My Wild Irish Rose*, was a biopic starring Dennis Morgan (1908–1994) as Irish-American tenor Chauncey Olcott (1858–1932) and featured a song of the same name written by Olcott that has become a standard. In the song, the wild Irish rose is both a flower given to the singer by a young woman and the young woman herself (see below).

My wild Irish Rose,
The sweetest flow'r that grows,
You may search ev'rywhere,
But none can compare
With my wild Irish Rose.

She is shy like a nodding flower and is found in a retiring place, but the singer hopes to win her heart:

Her glances are shy when e'er I pass by
The bower, where my true love grows;
And my one wish has been that some
*   day I may win*
The heart of my wild Irish Rose.

My Wild Irish Rose, CHAUNCEY OLCOTT, 1899

The 1979 movie *The Rose* was loosely based on the story of the rock and blues singer Janis Joplin (1943–1970), and starred Bette Midler as a vocalist named "The Rose." The title song, sung by Midler, but written by American lyricist Amanda McBroom, was a hit that won Midler a Grammy award. It presents the rose as a symbol of hope and love renewed.

French chanteuse Édith Piaf's hit song *La Vie en Rose* (which means something like "a rosy-colored view of life") was released in 1947 and became a big hit in the United States, where several major artists made cover versions, including Bing Crosby, Dean Martin, Louis Armstrong, and Grace Jones. The English version includes the lines below.

Just remember in the winter
Far beneath the bitter snow
Lies the seed that with the sun's love
In the spring becomes the rose.

The Rose, 1979

When you press me to your heart
And in a world apart
A world where roses bloom.

La Vie en Rose, 1947

THE ARTIST'S ROSE

A GREAT NUMBER OF ARTISTS in the twentieth and twenty-first centuries used images of roses in their work. American artist Georgia O'Keeffe (1887–1986) was celebrated for her large-scale images of flowers. She made more than 2,000 flower paintings in the course of her career.

O'Keefe's flower paintings included several of roses, such as *Abstraction White Rose* (1927), a big close-up of a white rose presented as a semi-abstract swirl of white interspersed with blush, gray-black shadow. Four years later she painted *Cow's Skull with Calico Roses*, an image of a cow's skull with two white calico roses—the sort of artificial flowers that are used in New Mexico to decorate graves.

In her large-scale paintings she wanted both to celebrate the beauty of flowers and play on their many cultural associations, while also—by making the images so large and striking—shocking people into really looking at them. She commented (see right):

A flower is relatively small. Everyone has many associations with a flower—the idea of flowers... nobody sees a flower—really—it is so small—we haven't time... So I said to myself—I'll paint what I see... but I'll paint it big and they will be surprised into taking time to look at it.

Exhibition of Oils and Pastels,
(January 22–March 17, 1939) GEORGIA O'KEEFFE

In the works of Salvador Dalí (1904–1989), the rose generally seems to symbolize youth and beauty, as is traditional. He used a recurring image of a woman with a bouquet of roses in place of her hair or her head. This appears in the sculpture *Homage to Fashion* (1971), *Alice in Wonderland* (1977), and *Femme à la Tête de Rose* (Woman with a Head of Roses) (1981), while in his sculpture *Woman of Time* (1973) a woman holds a rose aloft on a stem, with a melted clock (almost a Dalí trademark) draped over her arm. Dalí also painted a perfect red rose, apparently floating in the sky above a mostly bare yellow-brown (probably Spanish) landscape (see above) containing two figures in his striking *Meditative Rose* of 1958. Some interpret the two figures as lovers and the rose as a symbol of their love.

Modern Artist and the Roses

Other modern artists including the French Impressionist Claude Monet (1840–1926), German-Danish artist Emil Nolde (1867–1956), and the Swiss sculptor and draftsman, Alberto Giacometti (1901–1966) produced exquisite rose images. Monet lived much of his life at Giverny where he created a beautiful garden. His many flower and landscape paintings reflected his exploration of "controlled nature." Nolde painted a gorgeous oil painting named *Roses on the Path* in 1935, showing a bank of roses delivering a glorious splash of red color alongside a country path. In 1961, Giacometti made an excellent lithograph called *Bouquet of Roses*, showing the flowers in a vase on a tabletop.

American artist Alex Katz (b. 1927) has produced several intriguing images of roses and flowers. His striking rose painting of 2001, *Red Roses with Blue*, depicts red blooms and green leaves growing up a trellis or fence against a sky-blue image. His *White Roses* (2012) depicts ten white rose blooms, some in different states of opening, with green leaves and stems against a light blue background. His *Rose Bud* (2019) depicts a rosebud in two states of opening— more or less closed and more widely opened—against a greenish-blue background. All the pictures play with space and depth, and the apparent movement of the flowers across and out of the pictorial space is striking. Years earlier, in 1968, Katz compared painting flowers to painting images of people— he had often painted images of groups at cocktail parties; he said flowers are "all overlapping volumes" like the people in the group.

Another development in modern art was for the rose to become the work of art rather than be represented in it. In Dutch artist herman de vries's artwork (herman de vries, born 1931, spells his name in lowercase letters as a statement of his belief in equality) *108 pound rosa damascena* (2013–2015), which was part of the Dutch pavilion at the 56th Venice Biennale in 2015, thousands of rosebuds were laid out in a circle 13 feet (4 meters) in diameter on the floor. The room in which the artwork was created was filled with the perfume of the flowers, since as we have seen, *Rosa damascena* is grown and harvested for rose oil and perfume. de vries was an apprentice gardener for two years, beginning at age 18 in 1949 and in the 1960s worked at the Institute for Applied Biological Research in Nature at Arnhem in the Netherlands.

LEFT *The House among the Roses* (1925) shows Claude Monet's house at Giverny, France, as viewed through an avenue of roses. The dramatic pink and red roses are painted in the Impressionist style for which he was famous.

Towering Rose

German sculptor and conceptual artist Isa Genzken (born 1948) created a huge rose for her monumental artwork *Rose II* (2007). The flower in question is 36 feet (11 meters) tall and cast in painted aluminum and steel. It was first exhibited on the facade of the New Museum in New York City in 2010. *Rose II* is today in the Abby Aldrich Rockefeller Sculpture Garden at MoMa in Manhattan. It is called *Rose II* because Genzken had made a set of three earlier monumental roses in 1993 for art collector Frieder Burda from Baden-Baden—the German city famed for its rose growing and known as the country's rose capital. One of the original rose sculptures was installed in a private residence in Baden-Baden while a second was erected at Roppongi Hills in Tokyo.

Genzken also made the 26-feet (8-meter) tall *Rose III* from painted steel in 2016. It depicts a yellow rose and was modeled on a real rose, and stands in Zuccotti Park in Lower Manhattan. Some commentators have suggested a political reading for this installation, given the rose's symbolic association with democratic socialism and the fact that the park, a publicly accessible private property, was occupied in the 2011 Occupy Wall Street movements.

FASHIONABLE ROSES

The rose was one of the fashion industry's favorite flowers throughout the twentieth century. French fashion designer Paul Poiret (1879–1944), the "King of Fashion," who was so important that his influence on the world of couture has been likened to that of Spanish artist Pablo Picasso (1881–1973) on the world of fine art, picked the rose as the symbol of the haute couture fashion house he founded. He hired French illustrator and designer Paul Iribe (1883–1935) to create the celebrated rose outline. Designers including Christian Lacroix, Yves Saint Laurent, Domenico Dolce, and Stefano Gabbana have made repeated and beautiful use of the rose in their designs, for example, in Saint Laurent's rose-print dresses and Dolce & Gabbana's romantic rose-print jumpsuits.

In street art, the rose has proved an enduringly popular feature in tattoo designs.

Among the bustle of Lower Manhattan in New York, Isa Genzken's 26-feet (8-meter) tall *Rose III* sculpture reaches for the sky as a kind of floral-themed skyscraper.

THE POLITICAL ROSE

T HE RED ROSE BECAME ASSOCIATED with democratic socialism from the end of the nineteenth century onward. This appears to have been simply on account of the color red, linked to socialism since at least the time of the 1848 French Revolution. The French Socialist Party adopted the symbol of a rose in a fist in 1971 and the British Labour Party adopted the red rose as its symbol (replacing a red flag) for the 1987 election—according to some accounts, this was the decision of Peter Mandelson, though then party leader Neil Kinnock has gone on record saying it was his choice.

A non-violent resistance movement against the brutal government of the Third Reich in World War II took the name "the White Rose." White Rose members carried out a secret leafleting and graffiti campaign, calling for resistance against the government. Its leaders Hans (1918–1943) and Sophie Scholl (1921–1943),

The French Socialist Party logo

The color red is linked to symbols of socialism.

and Christoph Probst (1919–1943) were arrested by the Gestapo secret police and after a show trial were executed on February 22, 1943. Hans Scholl indicated under questioning that he was inspired to choose the name "White Rose" by a poem of that name by German Romantic poet Clemens Brentano (1778–1842), but some think this was a tall story to shield a bookseller called Josef Söhngen, who had given White Rose members a place to meet.

Another rose association with Leftist politics is through the slogan "Bread and Roses" and the poem of that title by American poet James Oppenheim (1882–1932). Oppenheim's poem was inspired by a 1911 article written by American women's suffrage activist Helen Todd (c. 1870–1953), who called for "bread for all and roses, too." Oppenheim's poem included the lines (see below):

The phrase "bread and roses" came to be widely used, suggesting that people needed both fair remuneration and decent conditions. Todd wrote:

The mothering element in the world and her vote will go toward helping forward the time when life's Bread, which is home, shelter and security, and the Roses of life, music, education, nature and books, shall be the heritage of every child that is born in the country, in the government of which she has a voice.

HELEN TODD, 1911

"Our days shall not be sweated from birth until life closes—
Hearts starve as well as bodies: Give us Bread, but give us Roses."

Bread and Roses, JAMES OPPENHEIM, 1911

COLOR MEANING

FLORIOGRAPHY, OR THE LANGUAGE OF FLOWERS, makes it possible to convey meaning without using written language. Tradition has it that the practice originated among women in Ottoman seraglios who were not permitted to communicate verbally. Roses have often been used to convey specific meanings according to their color. Where red roses symbolize love and passion, white roses stand for purity; pink roses reputedly symbolize happiness, and yellow roses friendship, though in German tradition the yellow rose symbolizes a declining love and infidelity.

Love and passion   Purity   Happiness   Friendship

LEFT A tarot card from the Rider-Waite deck, here the rose flower and the prickles represent strength.

Roses and their color symbolism also play a part in the Tarot deck used in divination and fortune-telling. Tarot cards, in use since at least the 1400s, were popular throughout the twentieth century and especially in the counterculture of the 1960s and 1970s; two decks commonly in use—the Rider-Waite deck and the Thoth deck—were issued in 1910 and 1969. The Fool card sometimes depicts a young man walking toward a precipice with a dog and a white rose—said to symbolize purity and freedom from lower desires. The Magician card generally shows a magus pointing to heaven and to earth, behind a table bearing symbols of the elements earth, air, fire, and water, and with roses and lilies in the foreground.

Because the rose unfolds from a bud to reveal its beauty, it was sometimes seen as a symbol of the revealing of wisdom through spiritual practices, while its fragrance suggested the sweetness of the wisdom. The Tarot Strength card usually shows a young woman with a lion; she often wears a crown or a belt of flowers, including roses; they reputedly represent the dual nature of strength—gentleness and ferocity, flower and prickles. The Death card features a skeleton or "Grim Reaper," often with a flag decorated with a white rose—symbolizing purity.

EPILOGUE:
THROUGH ROSE-COLORED GLASSES

As a desire to follow the advice of "do what you love," I started a rose garden design and maintenance business in Atlanta, Georgia, many years ago. I wanted to be surrounded by beautiful things such as plants—the flowers and trees that made me smile. Fortunately, I quickly became "the rose guy."

To begin this adventure, I turned to familiar varieties of roses that I knew and consulted the beautiful people in the rose world about what they were growing. I remember when my friend Pat Henry of Roses Unlimited in South Carolina and I were getting on an elevator at the Hotel of the District Rose Society. Pat was holding a potted rose, and when I asked, "which one is that?", she went into the friendliest of rants about the rose and told me all of its wonderful qualities. What struck me was the pure joy she had in describing how this rose was going to be in my garden. I loved that feeling. So, I went and heard more stories of roses and bought more… and more… and more. I acquired roses sold locally or online—mostly only deciding on the gorgeous picture of the rose. Often, there was the bottle(s) of fungicide and insecticide to go with them, and I abided. At the time, the purchase of plants and their chemical cohorts was a given. My clients demanded perfection in their roses, and I was going to do whatever was needed to deliver that.

After ten years, I moved on to New York City to be curator of the Peggy Rockefeller Rose Garden at the New York Botanical Garden. I was hired at a moment that the rose garden collection was to be revised and I inherited a rose garden that had been sprayed with chemicals for twenty years. After redesigning the collection, we reopened the garden in 2007. In 2008, we learned of a new local law that was to ban chemicals in the public spaces of New York. It made curating a world-class rose collection to the highest standards without chemicals a little daunting. So, I started over. I asked many people I knew if they had any roses that would do well in a hot, humid (disease-prone) climate that would work without chemical sprays. Some laughed and said good luck, a couple made suggestions, and a few emerged as having already started the work—I had a great platform if they had the roses.

As it turns out, there were some brilliant efforts toward disease resistance already in the works. In choosing disease-resistant varieties and implementing a healthy soils program, the Peggy Rockefeller Rose Garden had reduced the chemical applications by ninety-six percent by the time I left in 2014.

LEFT The Peggy Rockefeller Rose Garden was curated by Peter E. Kukielski 2006–2014.

THE INTENT BEHIND MODERN ROSE HYBRIDIZING

The word "hybridize" is defined as to crossbreed two different species or varieties (to create a new type). If I were a backyard rose hybridizer and wanted to create pink roses, I might take a red rose and cross it with a white rose to achieve a new pink tone. For the purposes of this story, you take this pink rose home, plant it, love it, and do every possible good thing for it, but then you find that over the season the roses that started off fine, got disease, never really had any fragrance, and died over the winter! I would argue that I was only going for pink; I never promised you anything else.

Perhaps the next season, you are wooed by gorgeous roses in a catalog. Again, you buy some, plant them, love them, and do every possible good thing for them, yet you get the same results. Now, you have two seasons of learning that you cannot grow roses. As rose consumers, we assume that the rose in that beautiful picture in that catalog has all of the appealing qualities we would ever want… including disease resistance, fragrance, and hardiness. However, sometimes all we end up with is Peter's non-performing pink rose or an equivalent.

When we looked at rose stories in this book, not once was there a mention of chemical applications that were being given to them at their moment in time. Were there bottles of fungicides needed to aid Cleopatra's rose efforts? The Greeks didn't have roses, violets, and insecticides for sale at the market! From over thirty-five million years of rose history, its proven resilience has already revealed its strength to find a deserved place in today's gardens. It is our modern manipulation of roses that has allowed for some of their incredible characteristics to get lost. Today, thank goodness, there have become some great efforts toward health, disease resistance, and fragrance!

BELOW The Floribunda rose 'Larissa' can flower from spring to fall and is highly disease resistant.

Stop Spraying to Discover Disease Resistance

The Kordes hybridizing house in Germany has been in the rose business since 1887. Such classic roses as 'Iceberg' ('Schneewittchen®'), 'Crimson Glory,' and 'Westerland' (KORwest) were early successes in the 1950s. The 1990s provided a challenge as Germany outlawed the use of chemicals for the entire country. The Kordes' efforts to produce healthy and disease-resistant roses were led by a (government-mandated) decision to stop the use of fungicides on their test fields. The result of this seemed to end in a pressure cooker of black spot in their growing areas—so strong, that nearly the whole field was defoliated. Kordes did find a few healthy varieties, which gave them a basis for new strains of rose breeding. Today, many of their new roses are multiple ADR-Trial (see page 245) award winners, revealing how strong and sturdy these roses have become without the use of chemicals.

Rosa 'Quietness,' hybridized by Dr. Buck, is tough and hardy as well as beautiful and fragrant.

Dr. Griffith Buck (1915–1991) was a horticulture professor at Iowa State University. He set out to breed hardy landscape roses and his roses became recognized worldwide for their winter toughness and low maintenance. From Professor Niels Hansen (1866–1950), Dr. Buck was given cuttings of *Rosa laxa* 'Semipalatinsk,' a species from Siberia belonging to the Cinnamomeae class, to aid with hardiness (zone 2b). The Kordes line also contributed to his breeding stock. Any rose from Siberia dictates tough and hardy!

Will Radler, a backyard hybridizer, wanted to eliminate the need for using harsh chemicals. He came up with the rose 'Knock Out®,' which has now been on the market for some years. 'Knock Out®,' and its siblings produce lots of color with healthy, maintenance-free plants—making them one of the top-selling roses. The seed parent of 'Knock Out®' was a seedling of the 'Carefree Beauty' variety (US Plant Pat. No. 4,225), 'Carefree Beauty' seedling × 'Razzle Dazzle' seedling. The toughness continues.

The shrub roses from David Austin Roses Ltd. have gorgeous blooms with a pleasant fragrance. These roses are tough and beautiful, which have made them a worldwide brand.

All these hybridization efforts are examples of individual intentions. From my over-the-top "pink only" roses, for instance, to Kordes and Radler focusing on disease resistance, Dr. Buck who produced roses that would be hardy in frigid climates, and David Austin who brings stunning flowers and fragrance.

For someone who wants to create a rose garden of their own, it is essential to understand the hybridization effort of the roses they choose. If you want a non-chemical garden, then choose varieties with a proven propensity toward disease resistance or specifically hybridized for disease resistance.

GLOBAL CHANGES FOR HEALTHIER ROSES

THE ROSE WORLD HAS CHANGED. There is a present conversation about healthy roses and the diminished use of chemicals in our gardens. This dialog is fantastic, and in some cases mandated: in Germany, New York City, and Ontario, Canada, the use of chemicals is prohibited. This trend seems likely to continue: the growing of roses without the use of chemicals gains strength each year as new healthy varieties are revealed.

Working with Mother Nature instead of doing battle with her.

BRAD JALBERT, 1995

Today, there are numerous people working hard toward healthier roses. Thank you to the big rose companies that are paying more attention to educating the consumer. I praise the small rose retailers for their endurance and passion and extend a thank you to people like Ping Lim with his Easy Elegance roses. Thanks to Brad Jalbert, who works with the motto quoted here (see above), and to Dr. David Zlesak, who promotes the education of black spot and disease resistance, along with hybridizing roses with superior hardiness and resistance.

I think of the Rose Rustlers—those determined folks who search for old roses that have survived the decades along back roads, old cemeteries, and homesteads. These roses prove their strong genetics and are being brought back into commerce for today's modern consumers to enjoy.

I also think of the trialing of roses around the world. The ADR trials in Germany are some of the most respected, and no chemical pesticides have been allowed since 1997. Award-winning roses have the attributes of disease resistance, hardiness, beauty, and growth habit. Many trials, such as The American Rose Trials for Sustainability (A.R.T.S.®) and the Earth-Kind® Rose Field Trials, are going on around the world, and we can only benefit from these future efforts. Thank you to all the people who are hybridizing and working toward a healthier future for roses.

Old Roses are New Again

Roses have taught us perseverance, strength, and beauty. Our observance of them (with the guidance of Mother Nature) is our education. The words of Farooq A. Sheikh (1948–2013) are appropriate here: "…I hate biochemistry because there is no chapter on roses." The world is such a beautiful place with all the roses in it, let us plant more roses and be students of the many things they teach us.

This book has told stories of the rose's resilience. Its millions of years of survival provide a foundation for its future. The timeless nature of the rose is safe because it is not a whim. *Rosa* will maintain its closeness in our days—both literally and symbolically. It will secure a place in that vase in our homes and our souls. Those who speak with an expression of the heart, do so with a rose.

He picked up one of Lorna's roses and set it in my lap. 'Here.' I picked it up and smelled it. He poked me in the shoulder. 'See what I mean? Thorns don't stop you from sniffing. Or putting them in a vase on the kitchen table. You work around them… Cause the rose is worth it… Think what you'd miss.'

Chasing Fireflies: A Novel of Discovery, CHARLES MARTIN, 2008

AFTERWORD:
PLANT A FEW SONGS ABOUT ROSES...

Plant roses for the future

Don't get me wrong, I'm a big tree hugger and avid consumer of farm to table, but because this is a book on roses, I get to put them first on the list. Roses will give you everything back and be a teacher of joy and beauty. I have often said that the only way to know roses is to grow roses; there are no classes one could take called "Roses 101." So, I take this opportunity to encourage the planting of a rose, so you may begin to learn their many ways of influence in our lives.

Plant for continuation

We can renew a commitment to plant roses, to inspire and dream, to experience the rose presently and for all future generations to come! Referencing Aristotle, the entirety of roses is always greater than the sum of its parts. Rose stories are told through millions of years, across continents, and are part of so many cultures and people—sometimes they have been fast and exuberant; other times rather slow and quiet. Never stopping, the encyclopedia of the rose is still being written. Its future is bright and based on its demonstrated celebration, perseverance, and resilience.

Plant roses for a healthier planet

In 2019, while standing in a new rose garden I designed at the Royal Botanical Gardens in Ontario, Canada, a woman asked me to lead her to a fragrant rose. She smelled it and found such joy that she then went around promoting her new aroma discovery, saying: "Come… smell this rose!" Julie's enthusiasm made for an immediate friendship. I proceeded to tell her that we created this new garden with the intention of balance. Although the roses were the stars, we surrounded them with companion plants that attract useful insects, which help to support the plants and ward off the bad insects. We work toward healthy soils where no chemicals or harsh fertilizers are used—only organics to feed the complex soil web, the vital cornerstone from which all plants grow.

Since this garden opened, turtles have laid their eggs next to rose bushes, and frogs have settled into blooms. Unexpected? Yes, but all signs that the intention of this garden is working. This commitment to health in the future presents a kinder, gentler rose garden. Let us aid Mother Nature, in her infinite wisdom, to do her thing. We can help her to start by encouraging the balance in the rose garden.

Plant rose memories

It was my Mom's Mom, Helen, who gave me my first memories of roses. Among so many things, she was a teacher of gardening. My grandmother and Papa's garden was my sanctuary: no matter what was going on in my life, I could escape to their oasis. There I would find the modern classics of the moment—'Peace,' 'Tropicana,' and the swooning smell of 'Mr. Lincoln'!

I didn't know it at the time, but their presence (both the roses and my grandparents) gave me comfort and taught me about caring, patience, and appreciation of nature. Papa used to go out and bury his morning banana peel next to each rose after he finished breakfast, teaching me valuable lessons on promoting healthy soils and natural fertilizers. These lessons continue to influence me, even now.

'Midas Touch'

Right before she died, my mother, Elizabeth, said to me: "If you ever see a yellow rose bush blooming, it means I am talking to you." One year to the day she died, this rose, 'Midas Touch,' was smiling in one of my gardens— "Hi, Mom!" She loved yellow roses, and I will always look for them and plant them for this very connection. Like the fragrance of the rose, just seeing one bloom can connect us to the past.

Plant roses for the joy

As I write, we are in the grip of the pandemic crisis of the COVID-19 virus. I imagine the year 2020 might be remembered this way. This horrible situation shakes every person in the world, and the news is grim. A recent article about a New York COVID-19 patient, Danny Burstein, describes his nurse helping with exercise for strengthening the lungs:

Breathing in through the nose she said: "Smell the roses!" And exhaling through the mouth she said: "Blow out the candles."

The Hollywood Reporter, April 13, 2020

It's times like these that I think of roses and the joy, life, and beauty they represent.

I am rooted in the memory of Leo Buscaglia (1924–1998). In *Living, Loving & Learning* he tells the story of taking a chocolate cake to a neighbor who just suffered the devastating loss of a family member. I am paraphrasing his words:

because no matter what, just like this cake, it is still a beautiful

world—like a perfect piece of chocolate cake, roses are part

of the deliciousness of this world.

Leo Buscaglia, 1982

I remember working at the New York Botanical Garden, where I was Curator of the Peggy Rockefeller Rose Garden. Every day, children would come to the garden, some witnessing and smelling a rose for the very first time. I knew that in some way their life changed at that very moment.

Plant roses because they are teachers

To finish, I quote from a favorite poem about roses:

[…] every summer

every rose

opened […]

to rise

in joyfulness, all their days.

Have I found any better teaching?

Not ever, not yet.

The Poet Visits the Museum of Fine Arts
from *Thirst* by Mary Oliver, 2006

The rose 'Peter's Joy,' was named for Peter E. Kukielski.

BIBLIOGRAPHY

Andersen, Hans Christian. Translated by Jean P. Hersholt. *Hans Christian Andersen's Complete Fairy Tales.* San Diego: Canterbury Classics, Printers Row Publishing, 2014.

Austin, David. *The English Roses.* London: Conran, 2017.

Austin, David. *The Rose.* London: Garden Art Press, 1998.

Beales, Peter. *Passion for Roses.* New York: Rizzoli Publishers, 2004.

Becker, Herman F. "The Fossil Record of the Genus Rosa." *Bulletin of the Torrey Botanical Club.* Vol. 90, No. 2, March–April 1963, (pp. 99–110).

Beveridge, Henry, ed. Translated by Alexander Rogers. *The Tuzuk-i-Jahangiri: Memoirs of Jahangir.* Pakistan: Sang-e-Meel Publications, 2001.

Blake, William. *The Complete Poems.* Alicia Ostriker, ed. London: Penguin, 1977.

Brown, Jane. *The Pursuit of Paradise: A Social History of Gardens and Gardening.* London: Harper Collins, 2000.

Bulwer Lytton, Sir Edward. *The Last Days of Pompeii.* London: Richard Bentley Publisher, 1834.

Burns, Robert. *The Complete Poems and Songs of Robert Burns.* Aberdeen: Waverley Press, 2011.

Burton, Sir Richard Francis, trans. *The Kama Sutra of Vatsyayana.* New York: Modern Library, 2002.

Calkin, Robert. "The Fragrance of Old Roses." Historicroses.org., 1999.

Calkin, Robert, and Stephan Jellinek. *Perfumery: Practice and Principles.* Hoboken: John Wiley & Sons, Inc., 1994.

Cameron, Mark. "A General Study of Minoan Frescoes with Particular Reference to Unpublished Wall Paintings from Knossos." Dissertation University of Newcastle upon Tyne, UK, 1974.

Campion, Thomas. *The Works of Thomas Campion: Complete Songs, Masques, and Treatises with a Selection of the Latin Verse.* Walter R. Davis and J. Max Patrick, eds. New York: W. W. Norton, 1970.

Coleridge, Samuel. William Keach, ed. *The Complete Poems of Samuel Taylor Coleridge.* London: Penguin, 1997.

Cruz, Juana Inés de la, et al. *Sor Juana Inés de la Cruz: Selected Works.* New York: W. W. Norton, 2016.

Day, Sonia. *The Untamed Garden: A Revealing Look at Our Love Affair with Plants.* Ontario: McClelland & Stewart, 2011.

Dey, S. C. *Fragrant Flowers for Homes and Gardens, Trade and Industry.* India: Abhinav Publications, 1996.

Downing, A. J., ed. *The Horticulturist and Journal of Rural Art and Rural Taste*, Vol II. New York: Luther Tucker, 1847–1848.

Eco, Umberto. *The Name of the Rose.* London: Vintage Classics, 2004.

Elliott, Brent. *RHS The Rose.* London: Welbeck Publishing, 2020.

Faulkner, William. *Collected Stories.* London: Vintage Classics, 2009.

Forbes, Robert J. *A Short History of the Art of Distillation.* Netherlands: Brill Publishing, 1970.

Fox, Rev. Matthew, and Lama Tsomo. *The Lotus & the Rose.* Montana: Namchak Publishing, 2018.

Freeman, Mara. *Grail Alchemy: Initiation in the Celtic Mystery Tradition.* New York: Destiny Books, 2014.

Geary, Patrick J., ed. *Readings in Medieval History.* Toronto: University of Toronto Press, 2015.

Gerard, John. *The Herbal or General History of Plants: The Complete 1633 Edition as Revised and Enlarged by Thomas Johnson Calla.* New York: Dover Publications, 2015.

Gordon, Jean. *Pageant of the Rose.* New York: Studio Publications, Inc. 1953.

Griffiths, Trevor. *The Book of Classic Old Roses.* London: Michael Joseph, 1988.

Grimm, Brothers. *The Complete Fairy Tales.* New York: Vintage Classics, 2007.

Harkness, Peter. *The Rose: An Illustrated History.* RHS. London: Firefly Books, 2003.

Heilmeyer, Marina. *The Language of Flowers: Symbols and Myths.* London: Prestel, 2006.

Henshaw, Victoria. *Urban Smellscapes: Understanding and Designing City Smell Environments.* New York: Routledge, 2013.

Hobhouse, Penelope. *The Story of Gardening.* London: Pavilion Books, 2019.

Iles, Linda. "The Isis Rose." rosamondpress.com/2012/10/06/the-isis-rose

Jackson, Shirley. "The Possibility of Evil," *Saturday Evening Post*, December 18, 1965.

Jashemski, Wilhelmina F., et al., eds. *Gardens of the Roman Empire.* Cambridge: Cambridge University Press, 2017.

Johnston, William M. *Encyclopedia of Monasticism.* New York: Routledge, 2015.

Keats, John. *John Keats: The Complete Poems.* London: Penguin, 1977.

Khan, Hazrat Inayat. *The Way of Illumination.* India: Motilal Banarsidass Publishers, 2011.

Kukielski, Peter E. *Roses Without Chemicals.* Portland: Timber Press, 2015.

Landsberg, Sylvia. *The Medieval Garden.* Toronto: University of Toronto Press, 2004.

Lane Fox, Robin. *Alexander the Great.* New York: Penguin, 2004.

Leffingwell, John C. *Rose (Rosa damascena).* Leffingwell & Assoc. Leffingwell.com/rose

Lovelace, Richard. *Poems of Richard Lovelace.* C. H. Wilkinson, ed. Oxford: Oxford University Press, 1953.

MacDougall, Elizabeth. *Medieval Gardens: History of Landscape Architecture Colloquium:* v. 9. Washington, D.C.: Dumbarton Oaks Research Library and Collection, 1986.

Macoboy, Stirling. *The Ultimate Rose Book.* New York: Harry N. Abrams, Inc. Publishers, 1993.

Moldenke, Harold N. *Medieval Flowers of the Madonna.* Ohio: University of Dayton, 1953.

Moore, Thomas. *A selection of Irish melodies with symphonies and accompaniments by Sir John Stevenson Mus. Doc; and characteristic words by Thomas Moore Esq.* Fifth number. London: Addison & Hodson, 1845.

Moore, Thomas. *The Poetical Works of Thomas Moore.* London: Bliss Sands & Co., 1897.

Newman, John Henry. *Meditations and Devotions.* Orleans: Paraclete Press, 2010.

Olcott, Frances Jenkins. *The Wonder Garden.* London: Pook Press, 2013.

Oliver, Mary. *Thirst.* Boston: Beacon Press, 2006.

Oppenheim, James. "Bread and Roses," *American Magazine,* December 1911.

Parsons, Samuel Browne. *The Rose: Its History, Poetry, Culture, and Classification.* New York: Wiley & Putnum, 1847.

Parsons, Samuel Browne. *Parsons on the Rose.* New York: Earl M. Coleman Enterprises, 1979.

Phillips, Roger and Martyn Rix. *The Quest for the Rose.* New York: Random House Inc., 1993.

Piesse, George William Septimus. *The Art of Perfumery: And Method of Obtaining the Odors of Plants.* Michigan: University of Michigan Library, 2004.

Plutarch. Translated and edited by J. Dryden and A. H. Clough. *Greek and Roman Lives.* New York: Dover Publications, 2005.

Potter, Jennifer. *The Rose, A True History.* London: Atlantic Books, 2010.

Potter, Jennifer. *Seven Flowers and How They Shaped Our World.* London: Atlantic Books, 2013.

Preston, Diana. *Taj Mahal: Passion and Genius at the Heart of the Moghul Empire.* New York: Walker & Company, 2008.

Reinarz, Jonathan. *Past Scents: Historical Perspective on Smell.* Illinois: University of Illinois Press, 2014.

Rumi. Translated and edited by Andrew Harvey. *Call to Love: In the Rose Garden with Rumi.* New York: Sterling Publishing, 2007.

Saint-Exupéry, Antoine de. Translated by Richard Howard. *The Little Prince.* New York: Houghton Mifflin Harcourt Publishing, 2000.

Sawer, John Charles. *Rhodologia: A Discourse on Roses, and the Odor of Rose.* Mishawaka: Palala Press, 2018.

Schieberle, Peter and Luigi Poisson. "Characterization of the Key Aroma Compounds in an American Bourbon Whisky by Quantitative Measurements, Aroma Recombination, and Omission Studies." *Journal of Agricultural and Food Chemistry.* Vol. 56, No. 14, 2008 (pp. 5820-826).

Scott, Walter. *Anne of Geierstein: or, The Maiden of the Mist.* Edinburgh: Cadell and Co., 1829.

Semple, Ellen Churchill. "Ancient Mediterranean Pleasure Gardens." *Geographical Review.* Vol. 19, No. 3, July 1929 (pp. 420–43).

Shabistari, Mahmud. *The Secret Rose Garden.* Michigan: Phanes Press, 2002.

Shakespeare, William. Stanley Wells, Gary Taylor, John Jowett, and William Montgomery, eds. *The Complete Works: The Oxford Shakespeare.* Oxford: Oxford University Press, 2005.

Shakespeare, William. Jonathan Bate and Eric Rasmussen, eds. *The RSC Shakespeare: The Complete Plays.* London: Palgrave Macmillan, 2007.

Shelley, Percy Bysshe. Zachary Leader and Michael O'Neill, eds. *The Major Works.* Oxford: Oxford University Press, 2009.

Shiner, Larry. *Art Scents: Exploring the Aesthetics of Smell and the Olfactory Arts.* Oxford: Oxford University Press, 2020.

Smith, Amanda. *The Modern Science of Smell.* ABC.net.au, 2014.

Smith, William, ed. *Dictionary of Greek and Roman Geography.* Boston: Little, Brown, and Company, 1870.

Stein, Gertrude. *Geography and Plays.* Boston: Four Seas Co., 1922.

Su, Tao, et al. "A Miocene Leaf Fossil Record of *Rosa* (*R. fortuita* n. sp.) from its Modern Diversity enter in SW China." *Palaeoworld,* 25, 2015, (pp. 104–115).

Theophrastus. Translated by Sir Arthur Hort. *Theophrastus: Enquiry Into Plants and Minor Works on Odours and Weather Signs (Books 1–5, Books 6–9).* Cambridge, Mass.: Harvard University Press, 1916.

Thompson, Dorothy B. and Ralph E. Griswold. *Garden Lore of Ancient Athens.* Oxford: Oxbow Books, 2013.

Todd, Helen. "Getting out the Vote: An Account of a Week's Automobile Campaign by Women Suffragists," *American Magazine,* September 1911.

Toynbee. J. M. C. *Death and Burial in the Roman World.* Baltimore: Johns Hopkins University Press, 1996.

Tucker, Arthur O. "Identification of the Rose, Sage, Iris, and Lily in the 'Blue Bird Fresco' from Knossos, Crete (ca. 1450 BCE)." *Economic Botany.* Vol. 58, No. 4, 2004 (pp. 733–36).

Urbani, Peter. *A Selection of Scots Songs, Vol. 1: Harmonized Improved With Simple, and Adapted Graces, Most Respectfully Dedicated to the Right Honourable the Countess of Balcarres.* London: Forgotten Books, 2018.

Villeneuve, Gabrielle-Suzanne Barbot de. Translated by J. R. Planché. *Madame de Villeneuve's Original Beauty and the Beast.* London: Pook Press, 2017.

Voragine, Jacobus de. Richard Hamer, ed. *The Golden Legend: Selections.* London: Penguin, 1998.

Webster, Richard. *Magical Symbols of Love & Romance.* Minnesota: Llewellyn Publications, 2007.

Williams, Henry T., ed. *The Horticulturist and Journal of Rural Art and Rural Taste.* New York: Henry T. Williams for the Library of the New York Botanical Garden, 1873.

Winston-Allen, Anne. *Stories of the Rose: The Making of the Rosary in the Middle Ages.* Pennsylvania: Pennsylvania State University Press, 1997.

Wordsworth, William. Stephen Gill, ed. *The Major Works.* Oxford: Oxford University Press, 2008.

INDEX

Biographies

PETER E. KUKIELSKI
is a garden designer and an acclaimed
horticulturist who was curator of the
award-winning Peggy Rockefeller Rose
Garden at the New York Botanical
Garden from 2006 to 2014.

CHARLES PHILLIPS
is the author of more than
thirty-five books, including
The Medieval Castle (2018).

JUDITH B. TANKARD
is an art historian specializing
in landscape history. She is the author
of ten books, including volumes on
Beatrix Farrand and Gertrude Jekyll.

Acknowledgments

I dedicate this book to my Grandmother Helen, who gently showed me the way of the rose and everything it had to teach even though I was too young to realize it.

I am grateful to everyone and everything in my life that has brought me to this point.

I am grateful to all who helped me hone this manuscript in all of its stages: at Bright Press—Sorrell, Abbie, and Jacqui; Beth for your editing; Jane Lanaway for your beautiful design;

Jean at Yale for your guidance; Charles Phillips for your contribution; Gaye and Alex for your gentle guidance, and in particular to Kate Duffy for your friendliness in everything!

I am grateful to my dear friends for your constant support and enthusiasm for my rose niche.

I am most grateful for my husband, Drew Hodges, for wrapping me in love and support to do what I love.

PETER E. KUKIELSKI

Picture Credits

The publisher would like to thank the following agencies for permission to publish copyright material.

Alamy: 7, Glenn Harper; 16L Album; 20R Matthew Taylor; 38, 188 The Natural History Museum; 42 Walker Art Library; 47, 181 Dinodia Photos; 48 The Reading Room; 53 Chris Hellier; 55T Pjr Studio; 55B Interfoto; 59 Artokoloro; 64 Falkesteinfoto; 77BL agefotostock; 99 Archivah; 105 Science History Images; 125 North Wind Picture Archives; 152 incamerastock; 162, 165 Old Images; 164 Hamza Khan; 162 and back cover Art Collection 2; 168 Florilegius; 171 Fabiano Sodi; 173 The History Collection; 191, 236 Les Archives Digitales; 198 Steffen Hauser/botanikfoto; 203 Picture Art Collection; 209 Granger Historical Picture Archive; 217 PAINTING; 219 Heritage Image Partnership ltd; 223 Collection Christophel; 224 Pictorial Press; 227L Everett Collection, inc.; 233 Richard Levine

Bridgeman Images: 67 Archives Charmet; 70; 79; 81 The Maas Gallery, London; 94 Accademia Italiana, London; 118 British Library; 229 Photo © Christie's Images

David Austin Roses www.davidaustinroses.com: 21R, 22R, 24R

Getty Images: 15 Paroli Galperti/REDA&CO/Universal Images Group; 62 DEA/G. Dagli Orti; 69, 154, 193 Fine Art Photographic; 73, 142, 145 Print Collector; 84 Sepia Times; 102 Heritage Images; 106 Thekla Clark; 110/111 DEA/G. Nimattalah; 115 Leemage; 127 Mondadori Portfolio; 136 DEA Picture Library; 174 PHAS; 176/177 DEA/Archivio J. Lange; 215 Staff

iStock: 45, 56, 141, 194

Peter E. Kukielski: 20L, 21L, 22L, 26L, 243, 248, 249

Shutterstock.com: 2, 3, 5, 13, 17, 18, 19, 24L, 31, 32, 35, 37, 39, 41, 50, 60, 61, 71, 76, 77T, 77R, 82, 86/87, 89, 104, 108, 130, 137, 144, 153, 157, 160TR, 161, 178, 179, 205, 220, 222, 225, 227R, 235, 239

STAATSBIBLIOTHEK ZU BERLIN Preussischer Kulturbesitz,Orientabteilung (Ms. or. oct. 1602, fol. 48v): 133

Suntory Holdings Ltd. 197

Wikimedia Commons
10 Redouté Rosa damascena https://clevelandart.org/art/1965.291; 14T, 27L Salicyna; 14 bottom Yuri75; 23R Jengod; 25 Gartenfrosch; 26R Kelvinsong; 27R, 207 A. Barra; 28 Mr. and Mrs. Charles G. Prasse Collection/Cleveland Museum of Art; 34 Michael Wolf; 74 Krzysztof Ziarnek, Kenraiz; 91 Georges Jansoone; 93 Cleveland Museum of Art https://clevelandart.org/art/1977.128; 97 Wiki PD-Art Sotheby's New York; 100 Wiki PD-Art Musée Jacquemart-André; 112, 212 Gift in the name of Warren H. Corning from his wife and children https://clevelandart.org/art/1959.17; 117 Wiki PD-Art; 120 Wiki PD-Art http://www.patriziasanvitale.com; 122 Wiki PD-Art The Yorck Project (2002) 10.000 Meisterwerke der Malerei (DVD-ROM), distributed by DIRECTMEDIA Publishing GmbH; 129, 230 Wiki PD-Art Museo Thyssen-Bornemisza, Madrid; 138 Library of Congress http://lccn.loc.gov/500496951; 146 Wiki PD-Art National Gallery;149 Wiki PD-Art Bens Culturais da Igreja; 159 Wiki PD-Art Bilkent University; 160 Wiki PD-Art; 182/183 Wiki PD-Art Dauerleihgabe der HypoVereinsbank, Member of UniCredit; 185 Wiki PD-Art Gift of Wallace and Wilhelmina Holladay; 186 Wiki PD-Art Google Cultural Institute; 201 Wiki PD-Art The Walter H. and Leonore Annenberg Collection, Gift of Walter H. and Leonore Annenberg, 1993, Bequest of Walter H. Annenberg, 2002; 211 Wiki PD-Art Museo Nacional del Prado; 234; 241 Andrey Korzon

All reasonable efforts have been made to obtain permission to reproduce copyright material. The publishers apologize for any errors or omissions and will gratefully incorporate any corrections in future reprints if notified.